Publisher's Note

More than twenty-five years after initial publication, we are proud to reissue Jesse Dilson's classic text, *The Abacus*, in its original form. While readers will no doubt find references to computer hardware and "electronic computers" anachronistic, we have decided not to alter them, since the binary system Dilson describes is still relevant to today's computer technology. As for the abacus itself, it remains untouched by the decades…and centuries.

THE
ABACUS

The World's First Computing System:

Where It Comes From, How It Works,

and How to Use It to Perform

Mathematical Feats Great and Small

JESSE DILSON

Drawings by Angela Pozzi

 ST. MARTIN'S GRIFFIN ✖ NEW YORK

www.stmartins.com

Library of Congress Cataloging-in-Publication Data

Dilson, Jesse.
    The abacus / Jesse Dilson.
        p. cm.
    ISBN-13: 978-0-312-10409-2
    ISBN-10: 0-312-10409-X
    1. Abacus. I. Title.

QA75.D5 1994
681'.14—dc20

93-43387
CIP

To William and Rose

The quotations at the chapter headings are all taken from what is perhaps the greatest book for young people ever written, Alice's Adventures in Wonderland, *by Lewis Carroll. Under his true name of Charles Lutwidge Dodgson, Carroll was himself a mathematics professor.*

Contents

THE ABACUS

The White Rabbit put on his spectacles. "Where shall I begin, please your Majesty?" he asked.

"Begin at the beginning," the King said very gravely, "and go on till you come to the end: then stop."

1. Ten is a round number

Mention the number 10, and people will say, "Now there's a nice round number."

And it is. It is around us all the time, in fact. Just look at your hands and you will find ten fingers. People are equipped from birth with two hands with five fingers on each, making a grand total of ten fingers per person.

What we have just written is known as a "mathematical statement." But such statements have to be demonstrated. And the way to demonstrate that each person has ten fingers is simply to make a count of them. Nothing is easier; you don't have to be an expert mathematician to count up to ten.

Of course, counting fingers is too simple. All the same, it is an idea important to learning to use an abacus— the idea of grouping by tens. A fancier name for this

13

idea is the "decimal system," a name for which the ancient Romans are responsible. Our word "decimal" comes from the Latin *decem*, meaning ten. The English word "abacus" is also taken directly from Latin; it was this name the ancient Romans applied to their type of computer.

The Romans were hardly the first to use a computer based on the decimal system. They borrowed their abacus from the Greeks, who called it the *abax*, and the Greeks, in turn, borrowed it from an even more ancient people. The chances are that the first men on earth, shivering in their clothing of animal skins in dank caves, hadn't as yet invented a really useful calculating device, but they knew about the decimal system just the same. They must have used their ten fingers for counting then, just as we sometimes use them for counting now. In fact, our word "digit" means both "finger" and "number."

Imagine a business deal between two cavemen. One enters his friend's cave with a bundle of tiger skins over his shoulder. He dumps them on the ground and says:

"Got six skins for you. I gave you seven skins the other day, if you remember, which means that you now have thirteen skins altogether."

"Wrong!" objects the other. "You gave me twelve skins altogether. Six and seven are twelve."

The first man looks at him scornfully. "That shows how ignorant you are," he says. "Look here." And he proceeds to count the sum on his fingers, proving to his fellow businessman that seven and six really are thirteen.

Using fingers to calculate with is good enough where a few tiger skins are involved. But what happens when something more plentiful must be counted, for example, clam shells, such as some American Indian tribes used for trade purposes? How could a wealthy man add, say, 87 shells to 926 shells? If he had to rely on finger-counting to get the answer, it would take him hours.

With some kind of calculating machine, like an abacus, he could do the job swiftly and easily.

"Can you do addition?" the White Queen asked. "What's one and one and one and one and one and one and one and one and one and one?"

"I don't know," said Alice. "I lost count."

2. The ancients

Look at the "ear" at the top right corner of this page. Can you answer the White Queen's question? Certainly, if you stop and count all the "one's" slowly. This is easy to do if the question is put in written form, as it is here. But when the question is spoken, at a normal speed, it isn't easy to keep up with, as Alice found.

Suppose, though, that the White Queen had merely held up three fingers and asked, "How many do you see?" Alice would not be stumped for a moment; she would need no more than a glance to say "Three!"

And that is one thing mathematicians mean when they speak of a "number sense": the ability to recognize a quantity at once without going to the trouble of counting it out. Most of us, looking at a group of four pebbles, for example, can recognize that there are four in the group. We may even recognize five. But for seven or more, we might have to do some counting.

Primitive cavemen probably had the same kind of number sense, but that was about the limit of their mathematics. Even their ability to count was not too strong. They may have had words in their spoken language for "two" or "three," or even for "twelve" and "thirteen." But the chances are that they had no words for quantities greater than that. They used expressions like "many" or "great" to describe large amounts simply because they could not count that high.

Although primitive man had a spoken language, he had not as yet invented a way of writing down his thoughts. The best he could do was draw pictures. We know that he was a clever artist because he left sketches in the caves of Spain and other parts of the world. But it was to take thousands of years before he could set down so good a record of his ideas that someone else could read it and understand exactly what he was trying to say. It was to take even longer before he could invent a way of counting.

But that did not stop him from making some needed calculations. Even before he found how to write numbers, he had a way of keeping records.

The cave dweller often kept sheep; he needed their skins for clothing, their milk and flesh for food, and their horns for musical instruments of a simple kind. Naturally, he hated to lose even one of these valuable animals. He must have found a way to count his sheep when he herded them into their pens at night, and to count them again when he let them out the next morning to graze.

How did this primitive man, who could barely count on his fingers, manage to keep track of a large herd of sheep? Perhaps he used a handful of pebbles. For every sheep he drove into the pen, he dropped one pebble into a pot; in the morning, he took one pebble out of the pot for every sheep he let out of the pen. If he found a pebble still in the pot when all the sheep had gone, he knew that either a wolf had made off with one of his animals or a neighbor cave-dweller had absently slipped it into his own flock. He would then look for the robber with fire in his eye. If, on the other hand, a sheep still remained in the pen after all the pebbles had been removed from the pot, he would know that he had somehow gotten richer overnight and would probably forget to mention it to his neighbor.

Thus, a bunch of pebbles may have been the earliest calculating machine. Our own language hints at that. If you look up the word "calculate" in the dictionary, you will find that it comes from the Latin word *calculus*, which means "pebble."

As time went on, primitive man left his cave and took to building huts, often setting them close together with a wall around them for protection against wolves and fierce, wandering tribes. The people in these small cities made the different articles that keep people comfortable: clothing, crockery, cookery pots, combs, even cosmetics for the ladies. What they could not use themselves they traded to others, swapping their products for things they needed. There was trade not only within the towns but between them, and with the growth of trade came the beginnings of a money system. And with money came a need for scribes and accountants, educated men who could not only read and write, but could also add, subtract, multiply, and divide. A simple pot of pebbles was not enough for these men who handled very large sums. They had to have a system of counting. But what kind of system did they use?

Babylonia, a rich, bustling city which flourished some five thousand years ago in Asia Minor, had a written language of wedge-shaped (or *cuneiform*) characters. Paper was unknown then, so the writing was done on soft clay tablets with a stylus that looked something like a modern carpenter's spike. Then the tablet with its curious signs was baked in a furnace for hardening. Some of these clay tablets still exist in various museums around the world; others are always being dug up by archaeologists, scientists who study them and try to understand their meaning.

The Babylonians used a system of numbers which also had the strange wedge shape of their letters. Their sign for one, which we write 1, was

ρ

—not much different from ours. The Babylonian accountant who wanted to write a four would simply scratch that many of these symbols on his tablet, thus:

ᛈᛈᛈᛈ

which looks like flying banners in a parade but is nothing like our number sign 4. For 10, they used the sign:

〈

Writing a number like 19 must have been a job in itself, for the ancient accountant had to scratch this into his tablet:

which, as you can see, is a 10 followed by nine 1's, thus making 19 altogether. The number 21 would be written

〈〈 ᛈ

With two 10's and a 1, the total is 21.

We can imagine some bright Babylonian inventing a simple calculating machine for himself. He would make up several clay tablets, each with a number traced on it, and make his calculations by taking away some of the tablets or adding some on. For example, if he wanted to subtract 8 from 19, he would first put together the proper tablets for the 19:

Then, taking eight of the smaller 1 tablets away, he would have left:

which he would at once recognize as eleven. A collection of such tablets must have been very handy for the Babylonian school-boy who had to do some side figuring and didn't want to spoil the big tablet he was using for his homework.

The ancient Egyptians, who built the great pyramids and left huge, splendid statues like the Sphinx for us to marvel at, had something the Babylonians did not have. It was a kind of paper known as *papyrus*, which they made from reeds growing in their broad river, the Nile.

It is certainly easier to write on a scroll of paper than

on a clay tablet, so the Egyptians had that much advantage over the Babylonians. But when it comes to number systems, there isn't much choice between those of the two peoples. The Egyptians had just about the same system as the Babylonians except that they used different signs. In that system:

| | was the sign for 1;
∩ | was the sign for 10;
℮ | was the sign for 100;
🦴 | was the sign for 1000;
🧍 | was the sign for 1,000,000

They had others, but these were the most often used. The first sign, just a vertical stroke, is not much different from ours. The next one was supposed to represent a heel bone, the third was a scroll, and the fourth a lotus flower, a plant the Egyptians loved. The last is meant to be a man holding up his arms in amazement. To the Egyptians, one million was a tremendous sum.

The Egyptians wrote, not as we do, from left to right, but in the opposite direction, from right to left. Their way of writing the number 123, for example, was

|||∩∩℮

The first symbol, the one at the extreme right, is the scroll. From the list above, we can see that it stands for 100. Next are two heel bones. They each amount to 10, and both together make 20. Add this to the 100 of the scroll, and we have a total of 120 so far. Finally, there are three 1 signs, so that the entire number reads 123— as we write it the modern way.

The Egyptians were careless about the order in which they wrote their numbers. And in their system, the order really didn't matter. Even if we write those same Egyptian signs this way:

$$\cap^{|}{}_{|} @^{|} \cap$$

we still have the 100 symbol, the two 10 symbols, and the three 1 symbols. And we still add them to get 123.

The Romans, who set up their great civilization a thousand years or so after the Egyptians, were a little fussier with their number system. The order in which a Roman number is written *does* make a difference.

Let's first look at the signs the Romans used for their numbers. Rather than work with pictures as the Egyptians did, they got their alphabet to do double duty—as letters for writing, and as numbers for reckoning. If these signs look familiar, they should: the English alphabet was borrowed from the Latin language of the Romans.

In this Roman system,

> I is the sign for 1;
>
> V is the sign for 5;
>
> X is the sign for 10;
>
> L is the sign for 50;
>
> C is the sign for 100;
>
> D is the sign for 500;
>
> M is the sign for 1,000.

We use the word "is" in the list just above rather than "was" because this numbering system did not die with the Roman Empire. We still use it, for old times' sake. As a matter of fact, you can find it on any dollar bill. If you look at the back of the bill, you will see, inside a circle at the left, a pyramid with this inscription at its base:

MDCCLXXVI

If you think this is a word, just try pronouncing it. It is definitely a number. And with our list to help us, we can figure out what the number is. Unlike the Egyptians, the Romans wrote from left to right; the first sign, then, is M, which is 1000 according to our list.

The next sign, D, is 500, which means that MD is 1500.

The next two C's are a hundred each, so that MDCC gives 1500 plus 200, or a total of 1700.

After that, we have the letter L, or 50. Adding this to the MDCC we have already gives us a total of 1750.

Now add on the two X's, which together amount to 20. This plus 1750 makes 1770.

Finally, there are the V and the I, which are 5 and 1, or 6. The entire number is therefore 1776.

So far, the Roman system seems to be just like the Egyptian; figuring out the value of MDCCLXXVI was pretty much like figuring out the value of

|||∩∩℮

But the Romans had one rule the Egyptians did not: when a sign of low value is placed *before* a sign of higher value, the smaller number is *subtracted* from the larger. In the numeral IX, for example, the I, which has a value of only 1, comes before the X, representing 10. So the value of the combination IX is 10 minus 1, or 9. The same rule applies to the number IV, which is 5 minus 1, or 4. But notice that VI is 6 because here the smaller number I comes *after* the large number V, which means that we *add* the 5 and 1.

Thus we can say that the order of the signs in the Roman system is important; it *does* matter where the signs are placed: VI is different from IV, XL (40) is different from LX (60), CM is 900 while MC is 1100, and so on.

How did the ancient Romans calculate with this strange system? Well, it wasn't easy, but there was no use grumbling about it because it was the best they had. Addition wasn't too bad; to add two numbers, you simply group together all the letters of the same value. For example, to add VII to VIII (7 to 8), you put the V's and I's in two sets: VVIIIII. We can see that this result is 15, for it has two V's, or 10, plus 5 I's, or 5. A Roman, however, wouldn't like the looks of this VVIIIII answer. He would first change the two V's to an X, knowing that two fives are equivalent to ten. Then he would change the IIIII, which is visually confusing, to a V. And his answer would be XV.

Subtraction was a bit tougher. Suppose you wanted to subtract XX from L. You couldn't do it without first changing the L into its equivalent number of X's; since L is 50, it is equivalent to XXXXX, or 5 tens. Now we can make the subtraction simply by removing two of those X's. Three X's are left, and so the result is XXX, or 30.

Don't ask what multiplication and division in the Roman system were like. They were dreadful. You will find Roman numerals on the cornerstones of buildings, showing the date they were built, and you may see them on the faces of old clocks or as chapter headings in old books, but nobody today uses them to calculate with. There is no need to; we have a much better system now.

3. The moderns

The modern system of numbers is called the Hindu-Arabic system because it was probably invented in India and came to the western world by way of Arabia. We say "probably" because nobody is really sure where, how, and by whom it was worked out.

One advantage of this system over the others that came before it is that with only ten signs it can express any number, no matter how large. Other systems—the Babylonian, Egyptian, and Roman—had only a few signs, but had to use these same signs over and over for some numbers, making calculation and even recognition difficult. The ancient Greek and Hebrew peoples used almost their entire alphabets for their number signs, which made things confusing too.

The ten signs of the modern decimal system are 1, 2,

3, 4, 5, 6, 7, 8, 9, and 0. They didn't always look as nice and neat as this. An old Spanish manuscript written about a thousand years ago, had digits like this:

$$I\ \zeta\ \zeta\ \gamma\ \gamma\ \gamma\ 6\ 7\ 8\ 9\ 0$$

In fact, every person in those days had his own way of writing them. It was not until much later that the number symbols got to their present forms.

When you really look into our decimal system, you notice some strange things about it. In the first place, it has *no single sign* for the number 10, although it is a decimal system based on that number. There is a different sign for every number up to nine—there is even a sign for nothing, the 0—but when we get to ten, we find that there are suddenly *two* signs, a 1 and a 0 placed side by side like this: 10. Why?

To find the answer to that question, let's look at this diagram:

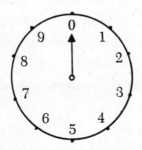

This is the dial face of a stop watch. Each of the numbers on it represents seconds of time.

Suppose this watch is in the hands of a timekeeper at a boxing bout. One of the boxers leads with his left. The other boxer, taking advantage of the fact that his opponent's guard is lowered, clips him with a round-house right to the jaw. Down goes the first man. The moment he hits the mat, the timekeeper presses the button on his stopwatch, and the hand begins to go around.

At the end of one second, the hand clicks over to the 1 position; at the end of two seconds, it clicks over to the 2 position; and so on. When ten seconds are up, the hand has completed a turn around the dial and comes to rest once more at the 0 mark.

The timekeeper then declares that the boxer reclining on the mat has been knocked out—kayoed, as the sports writers put it.

Let's suppose, now, that while all this was going on, the manager of the fallen fighter was deep in convers-ation with one of his friends at ringside. Unaware that his man has been knocked out, he rushes to the time-keeper in a rage.

"Why did you stop the fight?" he demands.

"Because your man was kayoed," says the timekeeper.

"Oh, he was, was he?" the manager says sarcastically. "Show me."

The timekeeper shows him his stopwatch with its hand pointing to the zero.

"That doesn't tell me anything," says the manager.

"Ah, but it does!" the timekeeper replies. He draws

a slip of paper from his pocket and writes a 0 on it. "That," he says, "is the zero the stopwatch hand is now pointing to. And that"—he writes a 1 to the left of the zero—"is the number of times the hand went completely around the dial. So the count is ten, and your man is out."

And, of course, he's right. By clicking through ten stops in going completely around the dial, the stopwatch has counted to 10—which illustrates how our modern Hindu-Arabic decimal system of numbering works. After going through ten units, the single digit it started with returns to 0, and a 1 is put before it. If the counting continues, the digit at the right goes from 0 to 1, from 1 to 2, and so on, until it reaches 9 again. Then it flops back to 0 while, at the same time, the digit to the left of it changes from 1 to 2. Thus, in any number with two digits—87, for example—the digit at the right stands for the number of units while the digit at the left stands for the number of tens.

Imagine that you are walking along a country road and a flying saucer floats down out of the blue. The hatch on top of the saucer opens silently, and out steps a strange creature. He hails you. It turns out that he's from the planet Jupiter and that he's come to Earth to find how far our science has advanced. As a starter, he asks you to explain our number system to him.

How would you explain it? You might begin this way: "In the first place, my Jupiterian friend, ours is a *positional* system. That means that if the position or order of the digits in a number changes, the value of the number changes."

The creature nods. "I get that," he says. "What's next?"

"Now," you continue, "if a digit is written all alone, it always represents units—and by units, I mean the things being counted; buttons, bananas, balloons, or whatever. For example, this digit"—you write the number 8 on a slip of paper—"represents eight units."

The Jupiterian writes this down in his notebook. "I understand so far. Please continue."

"Now, if I write another digit just to the left of that 8, the new digit is in the tens position. Look here. Suppose I write a 4 just to the left of the 8, like this: 48. How many tens do I have?"

"Four."

"And four tens make—"

"Forty," says the creature.

"Right," you say with a smile.

"But what about the 8? Does that still represent units?"

"Of course. And forty plus eight are forty-eight. So the sign '48' is read just that way."

"Good," he says, writing busily in his notebook. "Now, what would happen if we switch the two digits in 48 and make it 84 instead. Will it still be the same amount?"

"Oh no, never!" you object. "Absolutely not. Because now the 8 will be in the tens position, giving eighty, and the 4 will be in the units position, so that the number will read eighty-four. And this is certainly not the same as forty-eight. Remember, I said our system is positional, which means that we have to be careful where we put these digits."

"I am beginning to see daylight," says the Jupiterian. "Now what do you call the position to the left of the 4 in 48?"

"That's the hundreds place. If, let us say, we have a 9 to the left of the 4 to make the number 948, there will be nine hundreds, four tens, and eight units, making nine hundred forty-eight."

And you can see from the gleam in his eyes that, with his quick wit, he now understands the system completely.

"Aha!" he cries, "I see it now! Your system is based on ten; it is a decimal system. A digit alone is units; a digit just to the left of it is ten times units, or tens. A digit written just to the left of that is ten times tens, or hundreds. I suppose another digit to the left would be thousands?"

"Exactly," you say, "because a thousand is ten times ten times ten. If we put another digit to the left of 948, a 2, let's say, to make it 2948, the 2 would be in the thousands position, the 9 in the hundreds, the 4 in the—"

"Never mind the rest of it," the Jupiterian breaks in impatiently. "I'm a very busy spaceman, you know. By the way, do you have computers here on Earth?"

"Of all types. Maybe even more than you have up on Jupiter. For instance, there's a dandy little one called the abacus. You see . . ."

But before you start explaining the abacus, perhaps you'd better look at the next chapter.

"And if you take one from three hundred and sixty-five, what remains?"

"Three hundred and sixty-four, of course."

Humpty Dumpty looked doubtful. "I'd rather see that done on paper," he said.

4. Oriental know-how

We have so much paper around us all the time that we can't imagine a day when there were no scratch pads, no letter-paper or envelopes, no slips, chits, receipts, bills, food or chewing-gum wrappers, to say nothing of newspapers and magazines. Very often, if you want a scrap of paper to doodle on, or to figure out how many days are left till Christmas, all you need do is reach for it. And you can usually get it without stirring from your chair.

The same applies to pencils. Nobody would bother picking up a pencil stub he saw lying in the gutter; it wouldn't be worth the trouble.

But things weren't always that easy. As we saw, the ancient Babylonians had to go to all sorts of bother to make clay tablets. And there was no United Babylonian

Clay & Tablet Manufacturing Company, Inc. to turn it out for them by the yard as paper is made for us today by the mile. The Egyptians, of course, had their papyrus. But it isn't easy to change reeds growing on river banks into papyrus with nothing but the simplest kind of tools. Besides, only a handful of skilled men knew how to do it, and they were careful to keep it a secret from the common people.

Then how did the average man of those times do the little calculations everyone has to do at one time or another? We don't know for certain, but we can guess. He simply traced a series of horizontal lines on the ground like this:

The lowest line was for units, the next higher for tens, the third for hundreds, the fourth for thousands, and so on. Sometimes, because all these lines were a little dizzying, the man doing the calculating—we'll call him the "operator"—would draw an X mark on the thousand line to set it off from the others.

To form his numbers, the operator put pebbles on the proper lines. For example, the number 132 was set in by laying two pebbles on the first, or units line (because the last digit is 2 units), three pebbles on the second, or tens line (because the next-to-last digit is 3 tens), and one pebble on the third line up (because the first digit is 1 hundred). When the operator was finished with this job, his pebble pattern looked like this:

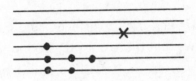

Now, suppose he wanted to add the number 61 to the 132 he already had. Since there is a unit of 1 in this number, he put one more pebble on the lowest, or unit, line. And since there are 6 tens, he put six more pebbles on the second line up, the line for tens. He wound up with this:

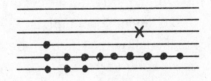

which is perfectly fine, except that he then had to count all those pebbles on the second line. That takes time. A wiser fellow would have placed just one additional pebble on the tens line, and a second pebble midway between the second and third line to represent a 5, like this:

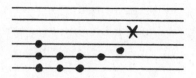

This result is easier to read. The single pebble on the third line shows that there is 1 hundred in the sum. There are four pebbles on the tens line, and since the pebble in between the second and third lines represents 5, we have a total of 9 tens, or 90. Add to that the three pebbles on the units line, and we see that the answer to 132 plus 61 is 193.

To an experienced operator, the answer is as plain as day. Once he has gotten his pebbles in their proper places, he can take in the final result at a glance. Yet he hasn't bothered to write down any numbers, he hasn't had to beg a piece of papyrus and a pen, nor has he gone to the trouble of baking any clay tablets. The only equipment he needs is a stick with which to draw lines in the ground and a few pebbles.

How do we know that such a device actually existed? Because a tablet with just such lines and just such an X-mark was dug up by archaeologists on the island of Salamis in Greece. That tablet is now in the Epigraphical Museum in Athens. It tells us that even before our present system of numbers was invented, people had a positional system in mind. We can see this more clearly if we turn the tablet counterclockwise, so that its lines are now vertical instead of horizontal. At the same time, we'll number those lines so we'll know which is which:

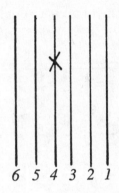

The tablet looked like a stepladder before; now it looks like a row of columns. But it's still the same. As before, the X marks the thousand line—or column—and from it we can tell that column one is still the units line, column two the tens line, column three the hundreds line, and so on. This is exactly the same as the positional system we explained in the last chapter; only then we talked about the position of digits while here we talk about the position of columns. We will see in a moment that the abacus follows the same system. In fact, we can say that this tablet of bygone ages is the grandfather of this book's hero.

We can see how a tablet like this would be pretty handy. The trouble with it, though, was that the pebbles used for counters were always getting lost, so someone decided that it would be a good idea to have rods instead of lines, with beads instead of pebbles sliding up and down the rods in the same way the beads of a necklace slide back and forth on their string. In that way, the beads are part of the instrument and are not likely to be mislaid:

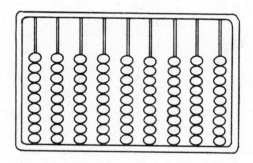

There are many opinions on the subject, but most historians think this early abacus was first created in Central Asia, somewhere in that vast stretch of land of the former Soviet Union. From there, it spread west to Europe, and east to China and its neighboring Oriental countries. Europeans did not care for it; they preferred doing their calculations on paper with pen and ink. The Chinese, on the other hand, took to it warmly; they realized they could do their computing much faster by sliding beads than by writing figures.

They liked it so well that they soon saw a way of improving it. Instead of using nine beads for each column, they used seven. Two of these seven had a value of five each, while the other five had a value of one each. Then, to make sure that the beads counting for five did not get mixed up with the beads counting for one, they put a crossbar in the upper part of the frame to separate the two types of bead from each other:

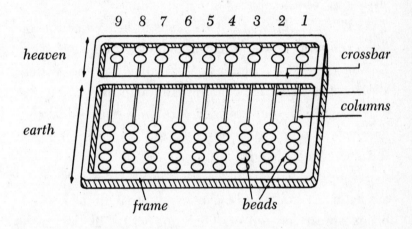

We can call this the "modern" Chinese abacus even though the Chinese have been using it for hundreds of years. It has a lot to offer. It is simple, it is run by hand and not by electricity, as so many of our computers are, it is easy to make (as we shall show later on), and it will operate for years—barring accidents—without having to be repaired. There are no parts to break down, which means that no spare parts are needed. Yet, in the hands of an expert, it can solve some problems just as fast or even *faster* than a modern electric computer (and we shall show that later on, too!).

As the sketch shows, the crossbar divides the frame into two areas, the smaller of the two on top, the larger below. A romantic people, the Chinese call the section above the crossbar "heaven," and that below it, "earth." Of course, we can give these two sections ordinary names like "the top part" and "the bottom part," but it's a shame to rob the abacus of its poetry.

As we said before, the two beads in heaven count for five each while those in earth count for one each. This is true of all the columns. Most abaci, as the sketch shows, have nine such columns, although there are quite a few with eleven, thirteen, or more. Naturally, the greater the number of columns, the bigger the figures the abacus can handle.

Before calculations begin, the operator lays the abacus flat on the table and makes certain that all the heaven beads are at the top of the frame while all the earth beads are at the bottom, just as they are in the drawing. This is known as "clearing" the abacus.

To bring numbers into calculation, the operator slips the proper beads up or down to the crossbar. The earth beads are moved up, of course, while the heaven beads are moved down. To take numbers *out* of calculation, the operator moves the beads from the crossbar back to the top or bottom of the frame.

As experts in the use of their own abacus, the Chinese have also worked out a neat way of handling it. They use the forefinger and thumb of the right hand for sliding the beads; the remaining fingers are kept out of use, curled in a loose fist. The forefinger has the job of controlling the beads in heaven, while the thumb does the same for the beads on earth. What you do with your left hand is not important; you can use it either to steady the abacus frame, which is usually a good idea, or you can keep it out of harm's way behind your back. This rule applies also to people who happen to be left-handed. As for the position the abacus should take while you're working it, the best one is flat on a table, or perhaps in your lap. But the frame should be flat. Tilting it may send all the beads in heaven sliding toward earth like so many fallen angels. This will not do your calculations any good, as any abacist can tell you.

And that is about as much information as anybody needs to start using this remarkable gadget. The rest depends on practice and common sense. Practice is needed to gain *speed*; common sense is needed for *accuracy*. If you devote a few minutes daily of idle time to the abacus—you can very easily work it while sprawling on the living-room couch—you will become an expert abacist in short order.

"I only took the regular course."

"What was that?" inquired Alice.

"Reeling and Writhing, of course, to be-
gin with," the Mock Turtle replied; "and
then the different branches of Arithmetic
—Ambition, Distraction, Uglification,
and Derision."

5. Reckoning on the rack

Now that we know how to hold and finger this Oriental
rack we call the abacus, let's get some exercise reckoning
on it. It's time to pick up your abacus and give it a test
run.

First exercise. A boy with 8 marbles gets into a game
with some boys, then leaves after winning 5 more.
He thinks he now has a total of 14 marbles. Is he right?

Aside from the fact that he showed sense by leaving
the game while he was ahead, let's see if he is an equally
sensible mathematician. Since he wants to know the total
of his original hoard plus his profit, his problem is one
of simple addition; he wants to add 8 to 5. This is child's
play for the abacus.

The first thing to do is make certain that the abacus is flat, and that all the heaven beads are up at the top of the frame, while all the earth beads are down at the bottom. With the abacus cleared, we feed in the first figure, the 8, by sliding one of the heaven beads in column one (the first column on the right) down to the crossbar, and sliding three of the earth beads in that same column up to the crossbar. Since one heaven bead has a value of five, bringing it down to the crossbar contributes 5 of the 8 we need; the remaining 3 is contributed by the three earth beads. The drawing here shows how this is done:

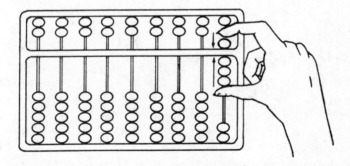

The picture also shows how the beginning abacist can gain speed. If the two actions of bringing down the heaven bead and bringing up the earth beads are done separately, one after the other, a split second is lost. By itself, this split second may not amount to much. But in a long operation, all the lost split seconds can delay the answer. The idea, then, is to avoid this loss of time by carrying out both actions at once. And that is done by following the Chinese system of using the forefinger for the heaven beads and the thumb for the earth beads. The two fingers pinch the beads together as a lobster grips things with his two-pronged claw.

With the 8 fed into the abacus, we have only to feed in the 5 to complete the problem. And we do that simply by flicking the heaven bead still at the top of the frame in column one down to the crossbar—with the fore-finger, of course.

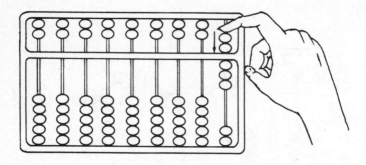

Now we can read out the result. The two heaven beads in the first column count for five each and so add up to 10. That, plus the three earth beads at the crossbar, make a total of 13. And so it turns out that the boy was wrong in adding up his marble wealth; he is actually poorer by one than he thought he was.

Let's see, though, if we can't show this answer in a more satisfying way. We know that the two heaven beads of this first column add up to 10. But we know also that the earth beads in the column just to the left of it (column two) count for ten units each, since this second column represents tens. In that sense, the Chinese abacus is just like the old Greek tablet-and-pebbles computer we talked about in the last chapter: the first column (or line) counts for units; the second counts for tens; the third counts for hundreds; and so on.

That being true, we can replace the two heaven beads in the first column by just one earth bead in the second. And this we do in one motion: we flick the two heaven beads up to the top of the frame with the forefinger, and slide the single earth bead of the second column up to the crossbar with the thumb. The bead pattern is then

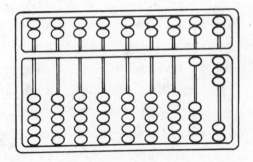

Here we have a result that is much easier to read out. Earlier, we read the result by adding 5 plus 5 plus 3. If we had to do that sort of thing every time we made a calculation, there would be no sense having an abacus in the first place. But by using this second method, we can

see the result at a glance: the column to the right has 3, the column to the left has 1. In the same way, the beads in these columns match the written number 13, which has a 3 to the right and a 1 to the left.

By the way, these expressions "feeding in" and "reading out" we have been using are part of the talk of electronic computer operators. Certainly, if these fellows have their special lingo, we can share it. We are in the same business, after all. The computer operator "feeds in" his figures by pressing buttons, turning cranks, or throwing switches; he "reads out" the result by looking at digits on a tape, a counter mechanism, or the glass face of what looks like a TV picture tube. The abacist feeds in his figures by adjusting his beads; he reads out his result by noting the positions of those beads. Both men are doing exactly the same thing.

Second exercise. Addition means combining two or more numbers to form a larger number; subtraction is the opposite, reducing one number by another. In the first exercise, we saw how the combining is done on the abacus. Let's see now how the reducing is done.

Suppose the boy of the first exercise gets into a second game with his 13 marbles. This time he isn't quite so lucky; up against a stronger player, he loses 9 of his thirteen. How many does he have left?

The problem here is to reduce 13 by taking out 9. The first step, as it always is with abacus computing, is to clear the abacus. The next is to feed in the larger number, 13. According to the rules, we should do this with the thumb alone, first shoving three earth beads of the first column up to the crossbar, then shoving one earth bead of the second column to the crossbar. But, for the sake of saving that split second we mentioned a few pages back, we'll use the thumb for the first column and the forefinger for the second. And this is shown in the drawing below:

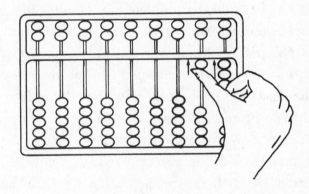

Now, how are we going to take 9 units from a units column having only 3? This is a puzzler, until we remember that the single earth bead in column two is worth 10 units in column one. If we take away that single bead, however, we'll be taking away one unit too many. So we add one earth bead in column one to those already at the crossbar, and flick the one bead in the second column back to join its fellows at the bottom of the frame. Since the only beads remaining at the crossbar are the four earth beads of column one, 4 is our answer.

Third exercise. Sooner or later, our boy loses all his wealth. His uncle promises that he will get him some more marbles if he can answer the following question: "If I buy six bags of marbles, four in each bag, how many marbles will there be altogether?" "Oh, that's easy," says the boy scornfully. And, picking up his abacus, he goes to work. In less than a minute, he has the answer. "Twenty-four!" he shouts.

The boy has solved a problem in multiplication. What his uncle wanted to know was, how much is 6 multiplied by 4? Or, putting it another way, how much are four sixes? The boy was smart enough to see that he could get the answer to this last question by adding four sixes together on his abacus. And this he did.

His first step (he knew that that is *always* the first step) was to clear the abacus. With that done, he fed the first 6 of his four sixes in by pinching one heaven and one earth bead together at the column one crossbar.

Since there was still another heaven bead remaining at the top of the frame in this column—to say nothing of four more earth beads at the bottom—our boy could feed in his second 6 by pinching that heaven bead and one of the earth beads together at the crossbar, like this:

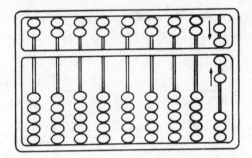

He now understood (especially if he read the earlier part of this chapter) that two heaven beads in the first column are the same as one earth bead in the second. So, with a single movement of thumb and forefinger, he flipped one earth bead in column two up to the crossbar and the two heaven beads of column one back up to the top of the frame.

With these two heaven beads back in the reserves, the boy now had fresh troops to throw into the battle. He fed in his third 6 by pinching together one of these beads and one of the three earth beads still in reserve, like this:

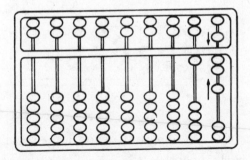

Then he fed in his fourth and final 6 by pinching in the remaining first-column heaven bead and the fourth of the earth beads in this fashion:

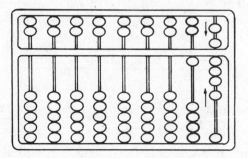

To get the answer with the least number of beads, he exchanged one earth bead of column two for the two heaven beads of column one just as he did before:

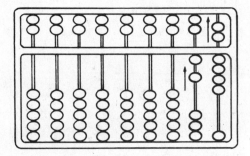

and wound up with two earth beads in the second column and four earth beads in the first. Since these were the only beads at the crossbar, they had to be the answer. He read it off promptly as 24.

We can't recommend this method for multiplying, say, 724 by 36, because that would mean feeding the figure 724 into the abacus 36 times. It would take much too long. But it is handy if you become absent-minded and forget how much 7 times 4 is, or whether 6 times 8 is 48 or 92. As for using the abacus to multiply larger numbers, we'll see how that is done in the last chapter.

Fourth exercise. Our hero pockets the 24 marbles his uncle gave him, and runs to show them off to the other seven members of his club. The club president, his eyes glued on the marbles in envy, tells him, "When we formed this club, Joe, we swore that anything any member has will be shared by all the other members. Right?" Our hero nods his head sadly; he knows what's coming. "Okay, then," the president goes on, "we'll divide your twenty-four marbles among the eight members in even lots." Since none of the members seem to know how to divide 24 by 8, our hero whips out his abacus—and how his friends open their eyes in amazement!—to work it out.

This, as the word "divide" tells us, is a problem in division. What the boys want to know is, how many 8's are there in 24. They can do that by subtracting groups of 8 from 24 until there are none left. By counting the groups they have subtracted, they get the answer.

Our hero begins to solve it by clearing the abacus. Then, he feeds the 24 into the first two columns:

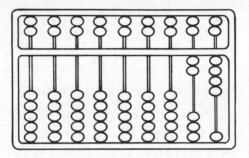

Now, he would like to take out the first group of 8. But since there are only four earth beads at the crossbar in column one, he can't do that directly. Our hero, however, has faced that problem before. He knows that one earth bead in column two is worth two heaven beads in column one, so he brings one of the beads in the second column down to the bottom of the frame and brings the two heaven beads of the first column to the crossbar:

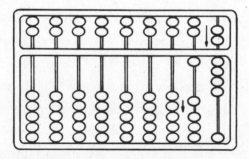

It's easy to remove the first 8-group now. He pushes one heaven bead in column one back up to the top of the frame, and three of the earth beads at the crossbar down to the bottom:

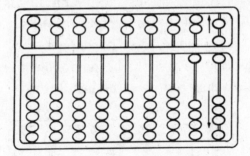

He remembers, now, that he has to keep a tally of the number of 8-groups he takes out of the original 24. So he brings one of the earth beads in the last column, way over at the other end of the abacus, up to the crossbar. He is using that column for his tally only.

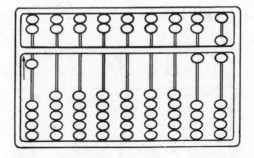

What he has remaining of the original 24 is a 1 in the second column and a 6 (one heaven and one earth bead) in the first. How does our boy get the second 8-group out? No problem there. If that one earth bead in column two is worth ten in column one, bringing it back to the bottom of the frame will subtract 10. But that is two more than our hero wants to subtract. So, at the same time he flips the bead in column two down to the bottom of the frame, he brings two earth beads in column one up to the crossbar, thus restoring the lost 2.

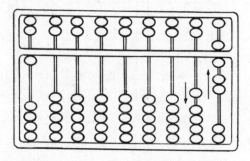

And, to indicate that he has taken another 8-group out of the original 24, he brings a second tally bead to the crossbar in the last column.

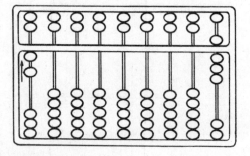

Looking back at column one now, our hero sees that he still has one heaven bead and three earth beads at the crossbar. This, of course, makes up his last 8-group. With a quick wipe of his forefinger and thumb, he whisks the heaven bead back up to the top of the frame and the three earth beads down to the bottom,

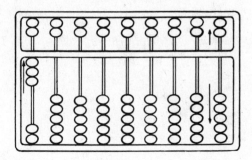

and adds another tally bead to the last column. Since nothing remains of the original 24, the three beads on the last column are the final answer. Every member of the club therefore gets 3 of our hero's hoard.

While none of the four exercises in this chapter are hard (it is in the last chapter that things get rough) they show how the abacus works. They also show that figuring with the abacus has these advantages over the usual pencil-and-paper kind of reckoning:

1. You don't have to write a single digit.

2. You don't have to ruin your eyes staring at figures and mistaking a 9 for a 7 or a 3 for a 5.

3. With a little practice, you can do your calculations in half the time you used to take.

4. It's more fun with the abacus.

There is still another advantage, but that will have to wait until the next chapter.

*"I know what you're thinking about,"
said Tweedledum; "but it isn't so, no-
how."*

*"Contrariwise," continued Tweedledee,
"if it was so, it might be; and if it were so,
it would be; but as it isn't, it ain't. That's
logic."*

6. Electrified abaci

Between the abacus, which looks like a toy, and the electronic computer, which certainly does not, there is a
family resemblance. There would have to be, since they
both do the same kind of job. But they are unlike in some
ways, too, just as brothers can be unlike.

Appearance is one of these ways. Most Chinese abaci
are small and delicate-looking, with the red and black
color scheme the Chinese are so fond of. An electronic
computer, on the other hand, is usually housed in a drab
metal box which may be large or small, depending on
the kind of problem the computer is meant to solve.
Some of them take up no more space than an average-
sized television set; others are big enough to fill entire
rooms.

Our nuclear submarines navigate by computer. The

rocket ships we have sent streaking to Mars and Venus are all equipped with computers of one kind or another. In fact, wherever lots of figures are handled—the Federal Reserve Bank, the New York Stock Exchange, the United States Post Office—computers are sure to exist.

If such computers were able to handle figures only, they would still be very much worth while having. But they can do more. They can use logic, much the way human beings can.

How do these electronic computers work? While they do the same kind of computing our abacus does, they do it in a different way. Abaci generally follow the decimal system which, as we know, is based on the number 10. While some electronic computers follow this system as well, many of them are built to work with what is known as the *binary* system. At least, it is known by that name to computer engineers, although the average man has probably never heard of it.

The binary system is really quite simple, once you get the hang of it.

If you look up the word "binary" in the dictionary, you will find that it means "double." And double, of course, means "two." That immediately tells us something about this new system: if the decimal system is based on the number 10, the binary system is based on the number 2. It tells us also that if there are ten digits in the decimal system—0, 1, 2, 3, 4, 5, 6, 7, 8, 9,—there must be only *two* digits—0 and 1—in binary.

We can see how the binary system works by remembering what we know about the decimal system. In the latter, a single digit stands for units. The largest of these single digits is 9. As soon as the units exceed 9, they are expressed as groups of 10 plus whatever units remain.

We have the same thing in binary. The largest of its two digits is 1. If we add another unit to this 1, it splits into two digits to become 10. Which means that, in binary, $1 + 1 = 10!$

But, if we were to read this last statement aloud, we could *not* say, "One plus one equals ten." One and one can never be ten. We still have to say "One and one equals two" because 10 in binary is the same as 2 in decimal.

Let's take a moment to remind ourselves what we explained before about our decimal system. We said that while a digit alone represents units, another digit written to the left of it represents tens, a third digit to the left of that represents ten times ten, or hundreds, a fourth digit to the left of that represents ten times ten times ten, or thousands, and so on. Bearing in mind that the binary system is based on two, we can simply substitute the word "two" wherever we used "ten" before. A digit alone, in binary, is still units. But a digit to the left of it represents *twos;* a third digit to the left of that represents two times two, or *fours;* a fourth digit to the left of that represents two times two times two, or *eights,* and so on.

Let's try a binary puzzle. What is the binary number 110 equivalent to, in decimal?

Here is the solution to that puzzle. The digit in the

first column to the right is 0, which means that there are no units. The digit to the left of that is a 1, and since that digit is in the twos position, it has a decimal value of 2. The third digit to the left is also a 1; it is in the fours position, and is therefore equal to 4 in decimal. Adding this 4 to the 2 gives a total value of 6. The binary number 110 is therefore equal to the decimal number of 6. The 0 wasn't added in, but simply holds open the units position.

If you're ready to try your hand at a second binary puzzle, see if you can figure out how much binary 101001 is in decimal. If it is done one step at a time, it turns out to be quite simple:

STEP A. The first digit on the right is a 1. This position represents units, so the value so far is 1 in decimal.

STEP B. The next digit to the left is 0. This is the twos position, but since there are none of these, there is nothing to add in. The total so far is still 1.

STEP C. The third digit is in the two-times two or four position. But this also is a zero, and four nothings are still nothing. The total so far remains at 1.

STEP D. Things are beginning to liven up. The fourth digit is a 1, and it occupies the two-times-two-times-two position. Three twos multiplied together give eight. Add that eight to the total of 1 we had in step C, and we have 9, in decimal numbers.

STEP E. The fifth digit is a 0 and, again, serves only as a place holder. Total is still 9.

STEP F. The final digit on the left is a 1. It is in

the position of two-times-two-times-two-times-two-times-two, or five 2's multiplied together. And that is 32 in decimal (2 times 2 is 4, 4 times 2 is 8, 8 times 2 is 16, 16 times 2 is 32). Add this 32 to the 9 we got in step E (and you can make this addition on the abacus, if you like) and we get the final result of 41.

So, the binary number 101001 is the same as decimal number 41. Hard to believe, isn't it?

If you don't like this result and demand a recount, we can show all these steps in a more entertaining way—in the form of a comic strip, for example. Here, the binary number of 101001 is represented by a squad of six knights in formation. The 1 knights are drawn in firm outline because they amount to something. The 0 knights, who just take up space, are drawn in a ghostly sort of way because they contribute nothing to the total score. The pennant each knight carries shows the value of his position: the fellow at the right carries a banner with a 1 on it because he is in the units position; the man next to him carries a 2 pennant because he in the twos position; the next man carries a 4 because he is in the two-times-two position; and so on down the line. As each knight is called forward, he shouts out how much he is worth, and the king—the knights' leader, of course—shouts the total number, in decimal, at each move.

Computer engineers, who use this binary system, have a language all their own. Since the expression "binary digit" is too much of a mouthful, they refer to the 1's and 0's of their counting system as "bits." This word "bit"

is made up of the first two letters in "binary" and the last letter in "digit." A combination of bits, like the 101001 we used as an example, is called a "word." When an engineer speaks of a six-bit word, he doesn't mean a dictionary item worth seventy-five cents, he means 101001, or 110011, or possibly 100001. As abacus operators, we are in the same professional family as these fellows who help build the marvelous machines we mentioned earlier. We should know something about their jargon.

We can see why a binary digit should be called a bit. But why call a collection of bits a *word*? There is a good reason for that, and we can understand it when we recall that computers are *logic* machines as well as calculators. Just as we think and reason in words, the computer thinks and reasons with words, made up of bits, fed into it by its operator.

But what do these computer engineers want with the binary system, anyway? Why can't they stick to the good old decimal system that everybody else has been working with for hundreds of years?

Actually, they know what they're doing. In just about every kind of electrical circuit imaginable, there is a switch of some kind. It may be the kind of switch all of us know about, embedded in practically every wall inside —and sometimes, outside—a house, the kind of switch that clicks up or down to turn on a light or turn it off. Or it may start a motor going or stop it. It might perhaps summon an operator on the telephone or cut her off. Just one average-sized city block may contain as many as a million switches, if we include not only those used for lights, motors, and telephones, but the switches on radios and television sets plus the ones *inside* those sets.

Not all these switches are the kind that are flipped by hand; some of them are turned automatically on and off by electric signals. But regardless of how they work, they all do the same thing: they either open a circuit or close it. When a switch opens a circuit, it breaks the path of the electric current; then a light goes out, a motor stops running, or a telephone goes dead. When the switch closes the circuit, the current resumes its flow; the light goes back on, the motor hums once more, and the person on the phone can continue talking.

Such simple switches have only two settings: open and closed, or off and on. Or, putting it in a mathematical way, 0 and 1.

And that is the secret of the electronic computer. The instrument is filled with tiny switches all busily operating. They are not the kind of switches ordinarily used for electric lights, motors, or telephones. They are much smaller (many of them about half the size of your thumbnail), are completely silent, and work much faster. The time they take to flip one way or the other is measured in *nanoseconds*, and a nanosecond is a thousandth of a millionth of a second. You might call it a super-split second. But every time one of these switches closes, it registers a 1; every time it opens, it registers a 0. Since these switches can have only two positions, it is perfectly natural that the computers they are in should use a counting system with only two digits.

It is hard to imagine how fleeting a nanosecond is, but think of it this way: if you were to blink your eyes once, as fast as you could, over a million nanoseconds would flash by while your eyelids were still stuck together. Once a computer gets going on a problem, it can solve it in a few such winks. Of course, the more complicated the problem, the longer the computer takes to work it out. But on occasion, the computer operator has to spend more time setting its different dials and push-buttons before we can get it started than the machine takes to find the answer.

The bigger computers are made up of several sections. One of them is the calculator, which contains most of the binary switches and does the actual figuring or thinking. Another is the memory, given that name because it stores information just as the human brain stores facts it has learned. Only, the computer memory keeps its information in the form of words—not the ordinary words used in speech, but groups of bits. Such pieces of information are stored in magnetic tape like that used in tape recorders, or in very tiny magnets known as ferrite cores. This information is always handy to be fed into the calculator when needed.

In other words, the computer works like a small and very simple brain. We say "simple" because, although it can work faster than the brains of its creators, it can deal with only a few types of problem while man's mind can ponder all the mysteries of nature, from the atom, so tiny that the eye cannot see it even with the most powerful microscopes, to galaxies so enormous that they can contain thousands of solar systems like ours.

The Chinese abacus, of course, is a much more humble instrument than the electronic computer. It has no memory device; it can do nothing but calculate. The problems it can handle are simple compared to those the computer wrestles with. It is not electrical, and hasn't a single switch. Yet it can still work as well in binary as it can in the decimal system.

What makes it useful for binary are the two beads in each column above the crossbar. Since the binary system deals with just two bits, 1 and 0, the heaven section of the Chinese abacus is nicely suited to this kind of reckoning. We must remember, however, that when we use the abacus for binary, the heaven beads have a value of *one* each. They can't count for five because there is no such number as five in binary.

Feeding bits into the abacus is much like feeding in decimal digits. (If you want to, you can call decimal digits "dits.") As we did with decimals, we work from right to left. The column to the extreme right is called the first column. The one just to the left of it is the second, the one to the left of that is the third, and so on. Bits are brought into the calculation by bringing heaven beads down to the crossbar, and are taken out of the calculation by sending them up to the top of the frame. We don't bother at all with the beads in the earth section, for we don't need them in binary work.

Before calculating begins, we clear the abacus by pushing all the heaven beads to the top of the frame. We are now ready to feed in bits. Suppose, to begin with, we set ourselves the problem of feeding in the "word" 101. The first bit, to the extreme right of this word, is a 1. We match this on the abacus by bringing one of the heaven beads in the first column—on the extreme right—down to the crossbar. The second bit in our word is a 0; this calls for no beads to be brought down to the crossbar, so we simply leave the second column empty. The third bit is another 1. We therefore bring one heaven bead in the third column down to the crossbar. And this is the bead pattern on the abacus:

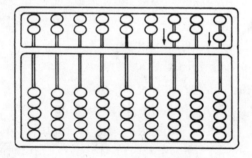

The bead positions in the abacus now match the word 101. In the same way, to feed in a word like 1011, we would slide one heaven bead down to the crossbar in the first and second columns, leave the beads in the third at the top of the frame, and bring one bead in the fourth down to the crossbar.

But the abacus was born for computing, not for registering numbers. So, suppose we try a simple binary calculation. Let's add the two-bit word 11 to another two-bit word, 10. We begin by feeding in either of the two—11, say. After clearing the abacus by sliding all the heaven beads to the top of the frame, we bring one heaven bead down to the crossbar in the first column, and another down to the crossbar in the second. That completes the feed-in of the 11 word:

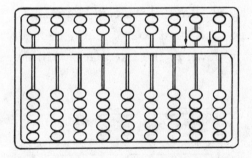

We now feed in the second word, 10. The 0 is no problem; it means only that the first column is to remain as is. As for the 1, we just slip the remaining heaven bead in the second column down to join its mate at the crossbar:

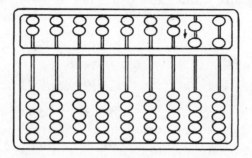

Now, with both words fed in, we should have the result. But, looking at the sketch above, we see something strange in it. The second column has two beads at the crossbar. It cannot be read as 2 because there is no such digit in binary. What we *can* do, however, is exchange these two beads in the second column for one bead in the third. So we bring these two beads back to the top of the frame, and send one of the heaven beads in the third column down to the crossbar. And this is the final result:

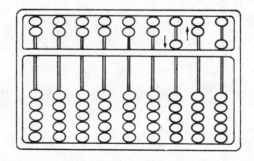

And that result is the word 101.

Now, what gives us the right to swap one bead of the third column for two of the second? We have every right. In binary, one bead in any one column is worth two beads in the column to the right of it. Our binary comic strip shows the same thing, with the pennant on each knight's lance flying a number which is twice that flown by the pennant before. And didn't we have a similar situation in the chapter just before this one, with decimal digits? We did, except that there, one column was worth *ten* times its neighbor, while here each column is *two* times the column next door to the right.

When we read the result of this calculation, we have to bear in mind that this is a *binary* result, not decimal. We can write it as 101, but we are committing a terrible mathematical crime if we pronounce it as one hundred and one. There is no such word as "hundred" in the language of binary. We might read it as "one oh one," but we must add "binary" just to make sure that the person we read it to doesn't mistake it for one hundred and one. In the same way, the two original words 11 and 10 must *not* be read eleven and ten, because "eleven" and "ten" have no meaning in binary.

As a simple exercise in binary subtraction, suppose we try taking the word 11 from the 101 result. We should get back 10; let's see if we do. After clearing the heaven part of the abacus, we feed in the 101. The first part of this subtraction is done by taking the 1 at the right of the 11 from the 1 at the right of 101. Nothing hard about that; we simply slide the single bead in the first column away from the crossbar back to the top of the frame. What we have left in the abacus is

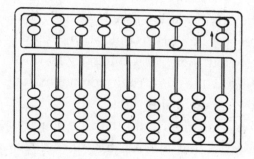

The second and final part of the problem is to take the 1 at the left of 11 from the 0 of the 101. In other words, we have to remove a bead from the second column. But how can you remove a bead when there is none at the crossbar in this column to begin with? We get around the difficulty this way: since the beads in column three are worth twice as much as the beads in column two, we can exchange the single bead in this third column for two beads in the second. So, we slide the bead in the third column back up to the top of the frame, and bring two heaven beads in the second column down to the crossbar, like this:

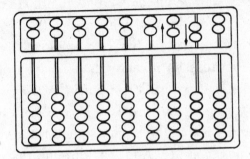

With two beads at the crossbar in the second column, we can easily make the subtraction of 1 by flicking one of the beads back up to the top of the frame. The result is

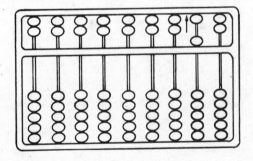

The answer is 10, one-oh binary.

The ancient Chinese who perfected the abacus had no more idea that his device could be used for binary calculations than that, some day in the distant future, a rocket (another Chinese invention, by the way) would be sent to the moon. In fact, it is very unlikely that the idea of a binary system so much as crossed his mind. He designed his instrument for decimal work only; it is just lucky chance that it is useful for binary as well.

But just because its inventor had no idea of binary doesn't mean that we can't use it in that way. Besides, our reason for doing so—a reason he was born too early to understand—is that the binary system is the system of the electronic computer. And it is through that remarkable machine that man may one day come to know his universe.

"I see you're admiring my little box," the Knight said in a friendly tone. "It's my own invention."

7. Venture into the mystic east

In one way, at least, an abacus is like a typewriter or a sewing machine. Once you find out what it's good for, you'll want to get one.

There is still another way the abacus, typewriter, and sewing machine resemble one another. To get any one of the three, you either have to buy it or make it yourself. But, where making a typewriter or sewing machine is concerned, even a highly skilled mechanic with a complete set of power tools would raise his eyebrows at the idea. Nor would the rest of us be likely to jump at the chance. As for buying either one, that takes a fairly large amount of cash.

That is where the abacus shines. If you're in the market for one, you'll find that it is far cheaper than either a sewing machine or a typewriter. A good, sturdy Chinese abacus with all the glamour of the mystic East about it (many of them are actually made in China) can be had for less than three dollars.

But before you rush out to the nearest odds-and-ends shop, listen to a piece of advice. Abaci sold in most stores are of two types, Chinese and Japanese. Whatever its color or the size of its beads, if an abacus has two heaven beads per column, it is Chinese. If it has only one heaven bead per column, it is Japanese.

We prefer the Chinese instrument. If someone else buys you the abacus as a gift, he ought to get the Chinese kind, with two beads in heaven. As the last chapter showed, those two heavenly beads are nicely suited to working with binary numbers. The Japanese abacus does not have that feature.

If you are the hardy and fearless type who prefers to make things himself, you can put together an elegant Chinese abacus without too much trouble. All you need, by way of materials, is a cigar box, a good, long stretch of what is known in the electrical trade as bell wire (about seven feet or so), and sixty-three beads. For tools, you should have a ruler, a small hammer, a nail slightly thicker than the wire, and a pair of wire cutters. This last item is part of almost everyone's tool kit, and so is not hard to borrow if you haven't got one yourself. Whatever you do, *don't* try to use the family scissors—unless, of course, you want to ruin them.

When you have your tools collected, go looking for a cigar box. That's the cheapest item you'll need; it will not cost you a cent. Most cigar dealers will let you have one or more. Choose a good-sized one with a hinged lid. If you have several that answer that description, pick the best-looking one.

Now, lay your ruler along the length of the box, on the side where the lid is hinged. Mark off ten evenly spaced divisions about one-half inch below the hinge with your pencil. If your box is ten inches long, the marks you make on it will be exactly one inch apart. If it is longer, the divisions will, of course, be larger; if shorter, the divisions will have to be smaller.

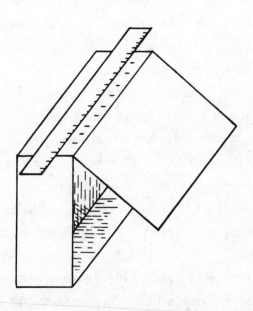

Write down the size of the divisions you have just made; you'll need it when you go shopping for the beads. These can be bought at any bargain store. It doesn't matter what color they are because you don't intend to wear them around your neck. Measured at their widest points over the hole through them, they should be about half one of the divisions you marked on the box, but no bigger. They may be smaller, of course, but not so small that you'll have trouble seeing them or sliding them back and forth when the abacus is finished.

When you buy the beads, take along a piece of the bell wire as a sample; those beads will later be strung on the wire, and you want to be certain that the holes through them are wide enough for the wire to slide through easily. An abacus whose beads move with difficulty is a slow abacus.

You are now ready to go to work. Using the hammer and nail, punch a hole right through each of the pencil marks you made on the box. When you're doing this, by the way, rest the side of the box being punched on some sturdy object—the corner of a workbench, for example— to avoid splitting it. Punch a second set of holes through the opposite side of the box. This second set should match the first exactly.

With your ruler, measure the distance between these opposite sides. Then, adding another inch and a half to this measurement, cut nine lengths of wire equal to the total. If the box measures eight inches, for example, the

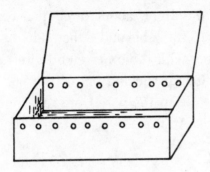

wire lengths you cut should be nine and one-half inches each.

First, string seven beads on each wire section. Now, poke one end of the wire through one of the holes in the side of the box, and the other end through the matching

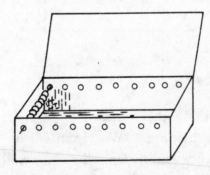

hole in the opposite side. Adjust the wire until you have equal lengths sticking out of either side of the box. If you've done this right, about three-quarters of an inch of wire tip will show beyond either side.

Do the same with the other eight wire lengths. Now, twist each wire tip sticking out of the box sides into a sort of knot to keep the wire from slipping through the punched hole. At this stage, your abacus should look like this:

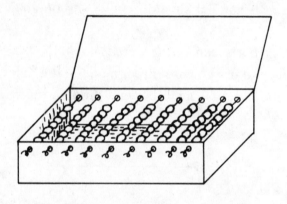

The only thing that remains to be done is to insert the partition dividing heaven from earth. Measure the length of the box exactly from the *inner* wall of one short side to the opposite wall, and cut an inch-wide strip of stiff cardboard of that length. The sort of cardboard from which shoe boxes are made will do nicely. Lay this strip across the wires in the box, and on it mark a little tick at each wire. Now, remove the strip and, with a sharp knife, cut slits to the depth of about one-half inch at each tick.

Finally, fit the strip into the box about one-third the distance from the hinged side to separate two beads from the remaining five of each wire. And here is a finished abacus built neatly into the box:

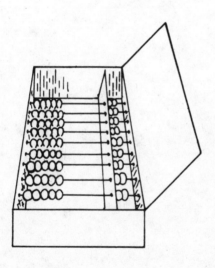

It is an instrument all your friends, even a Chinese scholar, will envy. It is portable—you need only tuck it under your arm to carry it anywhere without fear of damaging its insides—the lid will keep the box free of dust, and if you have chosen one of those beautifully decorated boxes the more expensive cigars are packed in, it will make a handsome piece for your room as well as a handy and practical computer.

Alice laughed. "There's no use trying,"
she said: "one can't believe impossible
things."

"I daresay you haven't had much prac-
tice," said the Queen. "When I was your
age, I always did it for half an hour a
day. Why, sometimes I've believed as
many as six impossible things before
breakfast . . ."

8. Math in the mystic east

An experienced Chinese abacist at work on his instrument has to be seen to be believed. The fingers of his left hand, clutching the abacus to keep it steady, remain still; but the fingers of his right race back and forth among the beads, sliding them swiftly up and down. Bowed intently over his work, he makes no sound. The beads, however, click merrily as they knock against each other, against the crossbar, or against the frame.

After just a few seconds, he raises his head and announces the result. To check it, the Westerner, to whom the abacus seems no more useful than a child's rattle, scribbles figures rapidly over his sheet of paper. But although he is just as good at arithmetic as the abacist, he finds that his calculations take at least twice as long. He throws down his pencil with a groan. "You're right," he says at last. "How did you get the answer so fast using that toy? It's impossible!"

The Chinese smiles gently. "It may seem impossible to you," he says, "but you must remember that we Chinese have been using this 'toy,' as you call it, for almost four hundred years. We should know how to handle it by this time."

"Four hundred years!" gasps the Westerner.

"That's right. Long before that, a crude kind of abacus came to China from Central Asia, and our ancestors improved it. This instrument you see here"—the Chinese points to his abacus—"is the result of their work. It was known to only a few people then, but some time later, in 1593, an early Chinese mathematician named Chen Ta-wei wrote a book called *Abacus Computation*. Only he didn't call his instrument an abacus; that is a Western name. He knew it by its Chinese name of *suanpan*, which means 'counting tray' in English. It was that book that made the abacus popular in China."

"Really!" the Westerner exclaims. "I didn't know that."

"Most people of the Western countries don't," the Chinese calmly replies.

"But it seems impossible that even in modern times these old customs are still kept up."

"Why shouldn't they, if the old customs are just as good and even better than the new? You saw a proof of that a moment ago when I worked out the problem faster on this ancient gadget than you could with your modern system."

The Westerner shakes his head and smiles. "It's true

that pencils are a fairly modern invention—a little over a hundred years old, I think—but the number system I used is just about as old as this Chen Ta-wei's book."

"That's true," the Chinese admits. "But when I speak of your modern system, I don't just mean a pencil. I mean the electric computer. That is a Western invention, and a very good one. But as fast as the computer is, our ancient abacus is even faster."

"Impossible! You're joking!"

"You use that word impossible too much. And I assure you that I am not joking. More than once, contests were held between the abacus and the computer, and the abacus won. For example, late in the 1940s a friendly race was held between a Japanese abacist and an American at the panel of an electric computer. And the Japanese won. Working on the same problem as the computer, he solved it quicker."

"But that's im— I mean, that's hard to believe!"

"Perhaps so," the Chinese nods, "but it's a fact just the same. I can give you names and dates. The contest was held on November 11, 1946, in the Takarazuka Theater in Tokyo, Japan, before an audience. The abacist was Mr. Kiyoshi Matsuzaki, a young worker in the Japanese Postal Administration, and the computer operator was an American soldier, Private Thomas Wood. Both men were given fifty problems in addition, subtraction, multiplication, and division. The Japanese won all the events except the multiplication. There, I must admit, he goofed."

"But that must have been just a freak," the Westerner protests. "It could never happen again."

"That's what *you* think. It happened again just three days later. A Chinese abacist named P. T. So, who was a student of banking at Columbia University in New York, beat an American opponent working with a computer. And they weren't given baby problems to do, either."

For a moment, the Westerner is silent. "Well," he says finally, shrugging his shoulders, "that was more than twenty years ago, and the machines we had then were not nearly as good as the ones developed later on."

"That's true. But even with better machines, the result was still the same whenever a really good abacist competed against them. For instance, a Chinese professor named Lee Kai-chen won a race against an electric computer in Seattle in 1959, and just to prove that it wasn't a fluke, did it again in New York. Some years later, contests like that were run in Taipei—which, as you know, is the capital of the Republic of China—and ended the same way. The abacus came out ahead."

"I still don't see how that could be possible."

"I've been wondering about it myself," the Chinese admits. "I think the reason is that the computer operator loses time setting the different knobs and switches on the machine while the abacist has no knobs or switches to worry about. But even so, it takes a lot of doing to outstrip a computer."

"I should think so," says the Westerner.

"Every Chinese child in grammar school is taught

arithmetic with the abacus," the Chinese explains. "Actually, it is the best way for children to learn it. All children like to play with the beads, and their interest in them grows when they discover that they are not only amusing but useful."

"I believe you're right," the Westerner says thoughtfully. "Is that kind of education true only in China?"

"Oh, no. The abacus is used in the schools of practically all the Oriental nations. And in the Soviet Union, too. In Japan, the abacus is called the *soroban*, and in Korea, it is known as the *chupam*. Russians call it the *abak*. Whatever its name, the abacus is just as important to the schoolboys and schoolgirls in these countries as their textbooks."

"And do the older pupils study it?"

"Certainly. As the student grows older, he is given harder problems to solve. If he wants to go on to college to study banking or business, he must take advanced courses in the use of the abacus. These courses are required; he must take them whether he wants to or not."

"But I don't understand," says the Westerner with a frown. "I'm sure that Japan and China have electric computers just as we have, and use them in their business offices. Then why should they spend so much time and money training their young people in the abacus?"

"For more reasons than one. In the first place, the abacus is much cheaper than the computer. An office, even a large one, can't afford computers for all its clerks, but it can certainly get abaci for them. Now, while it's true that the average abacist can't keep up with a machine—the men who won those contests I told you about were all top experts—he can certainly work far faster than the average pencil-and-paper clerk . . . . But there's another reason why the Oriental countries teach their young men and women to become good abacists."

"Oh? And what is that?"

"We are proud of our little invention, you see. You might say that it is part of our way of life. Many boys and girls in our country make a hobby of it and stage little competitions among themselves just as your Little League baseball teams compete against one another. And not only the youngsters, but the adults have their contests as well. For example, have you ever heard of the Asia Abacus Conference?"

The Westerner searches his memory. "No, I don't believe I have."

"I am not surprised," the Chinese says. "It is an organization with members in many Asiatic countries, all of them devoted to the abacus. Every so often, they hold a big international tournament. The way the members handle their instruments is a beautiful thing to see."

"It must be. Tell me, are the other Asiatic countries as handy with the abacus as you Chinese?"

"Oh, yes. The Japanese, especially. They adopted many Chinese customs, as you know, and one of them was the abacus. Their children start studying the *soroban* in the third grade, for about five hours a week. Students who show a talent for it are awarded special scholarships.

The Japanese—and we Chinese too, for that matter—issue licenses to expert abacists. To get one, the abacist must pass an examination. There are different ranks of expert; the higher the rank, the stiffer the examination."

"The Oriental people seem to depend a lot on these abaci," says the Westerner. "Would you feel lost without one?"

"Not really," returns the Chinese. "A good abacist can work his instrument mentally."

"Mentally! But that's remarkable!"

The Chinese bows. "Thank you for the compliment," he says. "But it is really not so remarkable when you remember that we have had a lot of practice. The trick is to have a mental image of the abacus, and work your fingers just as if you were handling the real thing. I know an expert who could amaze you with this kind of mental reckoning. He is rather like those master chess players who can defeat their opponents without even seeing the board."

"I am beginning to understand your respect for your *suanpan*."

"It is not only I who respect it. All Chinese do. As a matter of fact, the Republic of China has set May 10 aside as Abacus Day. Yes, it is very important to us. We have a little custom, you know. When a baby is born, an abacus and a writing brush are placed before him, one on each side. If the tot reaches for the brush, he will be a scholar; if he seizes the abacus, he is sure to be a merchant."

"Perhaps we Westerners should seize the abacus."

"You are. The United States alone imports something like 100,000 abaci a year. And some of your schools are beginning to teach it in their arithmetic classes."

"Well," says the Westerner after a silence, "you've talked me into it. You don't happen to have an abacus you can sell me, do you?"

"No. I have abaci, of course, but I wouldn't part with them for anything. And I have some really expensive ones, made of jade and ivory. But you can buy them in many stores at a much cheaper price, or you can make one yourself."

"I certainly intend to. Are they hard to learn?"

"Not in the least. Practice with one for about an hour a day, and within two weeks you will be fairly expert with it."

"I don't expect to win prizes, though," laughs the Westerner.

"No," the Chinese smiles in return, "but I'll bet a bag of litchi nuts you'll have fun trying."

"Who's been repeating all that hard stuff to you?"

"I read it in a book," said Alice . . .

9. Where things get rough

If you thought the four exercises of Chapter 5 were too simple and want to try something with more meat on it, you have come to the right place. This is the chapter where you can live dangerously.

The first "dangerous" operation is one in addition. Since the abacus is the instrument for businessmen, it is usually used for dollars-and-cents figuring. Our first exercise, then, will be to add this column of figures:

$73.22
146.67
29.93
2071.92

We will show the problem again at the top of each left hand page so that you can refer to it at a glance.

$73.22
146.67
29.93
2071.92

There are two ways of adding such a column on the abacus. One is to feed in the first horizontal row of digits, and follow it up with the second, third, and fourth rows. When that is done—and if it is done properly—we will get the right answer.

The second way is to make the addition vertically; that is, to feed in the four digits of the first column on the right, the four digits of the second column, the four digits of the third, the four digits of the fourth, the two digits of the fifth, and the lone digit of the sixth.

Personally, we prefer the second method. It may be the slower of the two, but it is easier to handle. However, we won't play favorites. Both methods will be explained, and in step-by-step fashion, through drawings.

The horizontal method. In the first or horizontal method, we begin by feeding in the top row of digits, starting with the 2 at the right and ending with the 7 at the left. These digits are fed into the first, second, third, and fourth columns of the abacus. After we have done this, the abacus beads have this pattern:

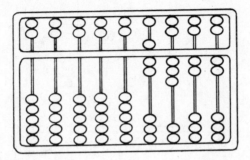

The next step is to feed in the digits of the second row, again from right to left. We begin with the 7. Since this is the digit to the extreme right in that row, we pinch one heaven and two earth beads into the extreme right column of the abacus, column one:

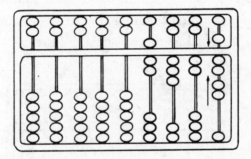

The next digit to be fed in is a 6, and this goes into the second column. We simply pinch one heaven and one earth bead together at the crossbar:

$73.22
146.67
29.93
2071.92
——————

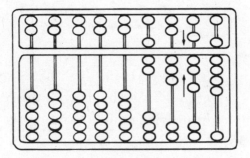

We now come to another 6. This is handled in the same way: we pinch together one heaven and one earth bead at the crossbar in the third column:

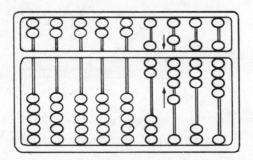

The next digit to be fed in is a 4, and it should be put into the fourth column of the abacus. But, looking at this fourth column, we see that there are only three earth beads at the bottom of the frame. How then do we get the 4 we need?

We can do it with this bright idea: we add 10 and subtract 6 to give us the 4 we need. Since one bead in the fifth column is worth ten of its equivalent in the fourth, bringing one earth bead in column five up to the crossbar adds 10 to column four. To subtract 6 from this fourth column, we simply flip the one heaven bead at the crossbar up to the top of the frame and one of the two earth beads at the crossbar in this column down to the bottom:

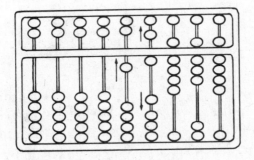

The last digit in the second row is a 1. No problem there. We simply flip one earth bead up to the crossbar in

> $73.22
> 146.67
> 29.93
> 2071.92

the fifth column to join the one we just put there. We now have two earth beads at this fifth column crossbar:

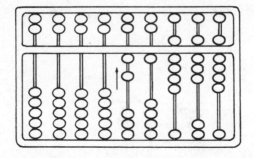

That finishes row two. We now tackle row three. The first digit to the right in this row is a 3, which means that we feed it into the first column. This we do by the same "bright idea" of adding 10 and subtracting 7. To add 10 to this first column, we bring one earth bead of column *two* up to the crossbar. To subtract 7 from the first column, we wipe one heaven and two earth beads at the crossbar to the top and bottom of the frame:

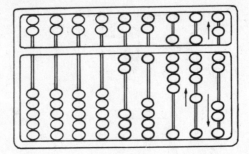

The second digit in the row is a 9. Since 9 is 10 minus 1, we can get this 9 into column two simply by bringing one earth bead in column *three* up to the crossbar while sending one of the earth beads at the crossbar in column *two* to the bottom of the frame:

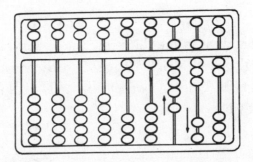

We now come to another 9, this one to be put into the third column. Again: We bring one earth bead to the crossbar in column *four* and remove one of the earth beads from the crossbar in column three:

$73.22
146.67
29.93
2071.92

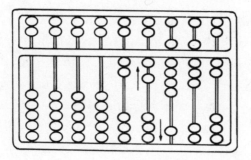

The final digit in the third row is a 2. Since this is the fourth digit from the right, it goes into the fourth column of the abacus. Adding two earth beads in this column to the crossbar, we get:

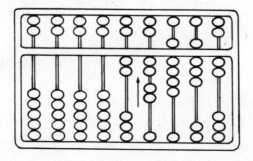

Now for the fourth and last row. The first digit on the right is a 2. We bring two earth beads in the first column of the abacus up to the crossbar. That gives us

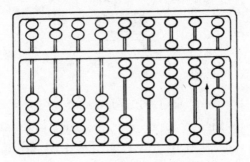

The next digit is a 9. Again, since this 9 must be put into the second column, we raise one of the earth beads in the *third* column to the crossbar and lower one of the earth beads in the second column to the bottom of the frame.

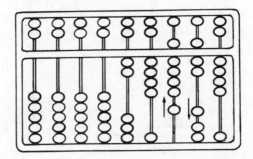

Continuing to the left in this fourth row of digits, we come to a 1. This must be fed into column three of the abacus. Looking at this column we see that it holds a

$73.22
146.67
29.93
2071.92

value of 10—one heaven bead, counting for five, and five earth beads for another five. And that gives us another idea: we can feed in the 1 we need by adding 10 and subtracting 9. So, we first bring one earth bead up to the crossbar in the *fourth* column. Then we flip four of the earth beads at the crossbar in the third column down to the bottom of the frame, and the heaven bead at the crossbar to the top of the frame:

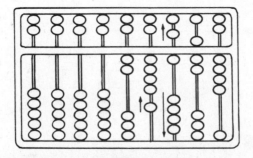

The next digit is a 7. This must be fed into the fourth column of the abacus since it is the fourth digit from the right. We get it in by adding 10 and subtracting 3. And that means bringing one earth bead in the *fifth* column

up to the crossbar and three earth beads in the fourth
column down to the bottom of the frame.

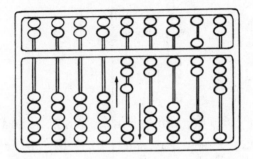

Now we have a 0, to be fed into column five. Adding
a zero to this fifth column means doing nothing to it. So
we pass on to the next column, the sixth.

This calls for the addition of a 2, the final digit to be
fed in. And this is very easy; we just slip two earth beads
in the sixth column up to the crossbar:

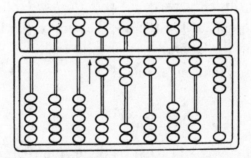

And we have the result. If we want to write it down,
we can do it by reading the abacus either from right to

$73.22
146.67
29.93
2071.92
———————

left—that is, by beginning with column one and finishing with column six—or from left to right, beginning with column six and ending with column one. Suppose we do it the first way. Since column one has four earth beads, it corresponds to the digit 4. Column two has one heaven bead, counting for 5, and two earth beads for a total value of 7. Column three has a single earth bead, so its value is 1. Column four has two earth beads; its value is 2. Column five, with three earth beads, is 3. And, finally, column six has two earth beads for a value of 2. Of course, in making this tally, we count *only the beads at the cross-bar*. The others at the top and bottom of the frame are out of the calculation.

So we see that our result is

232174

But something still must be done. We remember that the figures we added were dollars and cents. So we put the finishing touches to our answer by writing a dollar sign in front of the 2, and a decimal point between the

1 and 7 to indicate that the 74 to the right of the decimal point is cents:

$$\$2321.74$$

And, if you like, you can put a comma after the 2 to clarify immediately that the 2 represents thousands:

$$\$2,321.74$$

This makes it a little easier to read, and we read it as two thousand, three hundred, twenty-one dollars and seventy-four cents.

The vertical method. To show how the vertical method is used, we'll add the same figures with it, as they appear at the top of the page.

And this is how we'll go about it. First, we'll feed all the digits at the extreme right into the first column of the abacus. Those digits, as we can see, are 2, 7, 3, 2. Then we'll feed the digits just to the left of them (2, 6, 9, 9) into the second abacus column. We'll continue in that fashion until we've worked our way down through every column. Then—we hope—we'll have the right answer. Of course, the answer we get by this method must be the same as the one we got by the first. Let's see if it is.

The very first step, of course, is to clear the abacus.

With that done, we feed in the top digit to the extreme right—2. We bring two earth beads in the first column of the abacus up to the crossbar:

$73.22
146.67
29.93
2071.92

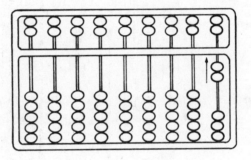

The next digit is the one just below, a 7. To feed that
in, we pinch one heaven bead and two earth beads of
the first column in at the crossbar:

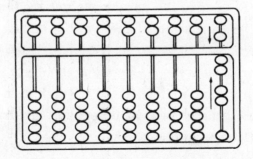

So far, so good. The next job is to feed in a 3. Here we can use our "bright idea" of adding 10 and subtracting 7 to get the 3 we want. We add the 10 by raising one earth bead in the *second* column to the crossbar and subtract the 7 by returning one heaven and two earth beads from the crossbar in the first column to the top and bottom of the frame, like so:

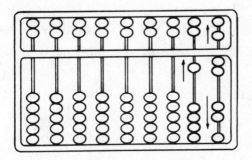

The next digit to feed in is a 2. We simply bring two earth beads in the first column up to the crossbar:

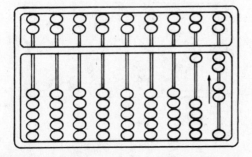

That finishes our first column of digits. We are now ready for the second. Since the columns in the abacus

$73.22
146.67
29.93
2071.92
————

match the columns of digits, we shall feed these second column digits into the second column of the abacus. Here we go:

The first digit of the second column is a 2. We therefore bring two earth beads in the second abacus column up to the crossbar to join the one already there:

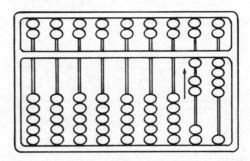

Below that 2 is a 6. We bring this into the second column of the abacus by pinching together one heaven and one earth bead:

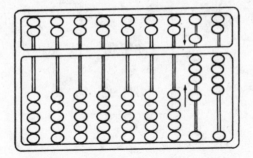

Below that 6 is a 9. This calls for our "bright idea." We bring one earth bead in the *third* column up to the crossbar to add 10, and drop one earth bead in the second column from the crossbar to the bottom of the frame to subtract 1:

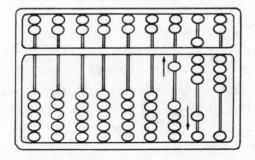

Now we come to another 9. We do just as we did before:

$73.22
146.67
29.93
2071.92

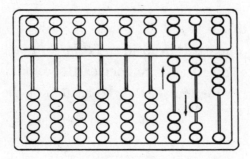

That takes care of the digits in column two. We now turn to column three. The top digit here is a 3. We just slide three earth beads up to the crossbar to join the two already there:

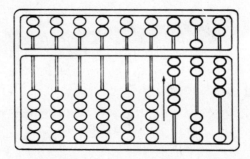

The digit under the 3 is a 6. Again: We bring one earth bead in the *fourth* column up to the crossbar, and send four of the earth beads at the crossbar in column three down to the bottom of the frame:

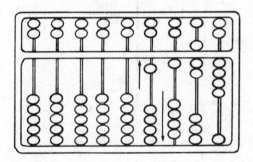

We come now to a 9. We bring one earth bead in column four up to the crossbar and retire the one remaining earth bead in column three to the bottom of the frame:

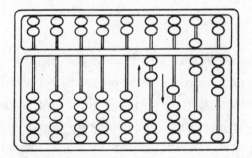

The last digit in this third column is a 1. So, in column three of the abacus, we bring that one earth bead back to the crossbar where it was before:

```
          $73.22
          146.67
           29.93
         2071.92
         ────────
```

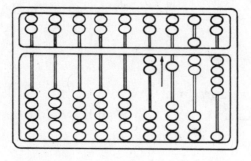

This may seem silly, this business of slipping the same bead back and fourth, but if you try to avoid it by doing two operations at once, you'll come to grief and wind up with the wrong answer. An expert might manage it, but the beginner should adopt the motto "one step at a time."

We come now to column four, in which the top digit is a 7. We feed it into the fourth column of the abacus by pinching one heaven bead and two earth beads together with the two beads already at the crossbar:

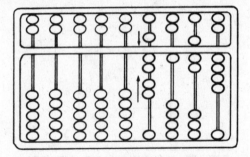

The next digit below that is a 4. A "bright idea" is called for here. We bring one earth bead in the *fifth* column up to the crossbar, and take one heaven and one earth bead away from the crossbar in column four:

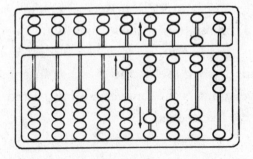

Beneath that 4 is a 2. This is fed into the fourth column the easy way, by bringing two more earth beads up to the crossbar.

$73.22
146.67
29.93
2071.92
———

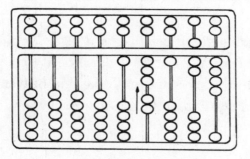

The last number in the fourth digit column is a 7. We feed it into column four of the abacus by raising one earth bead in column *five* to the crossbar and removing three earth beads from those at the crossbar in column four. This, as we know, is the same as adding 10 and subtracting 3 to give the needed 7.

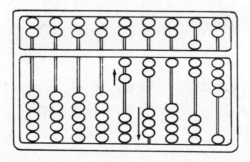

And that takes care of the fourth column of digits. Now for column five. Glancing at the figures above, we see that this column has only two digits, a 1 and a 0. We can forget about the 0 except to note the position it holds. The 1, however, has to be fed into column five of the abacus. To do that, we simply bring one earth bead up to the crossbar to join the two already there.

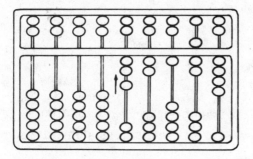

The sixth column of digits has just a 2. So, in this column of the abacus, we bring two earth beads up to the crossbar:

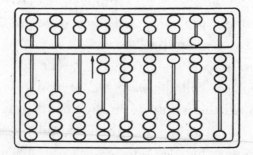

That completes the columns and the problem. Reading the result from the picture just above—or from your own abacus, if you've been working the problem along with us—you can see that it is

232174

or, in dollars and cents,

$2,321.74

which is exactly what we had before.

Now that we've cut our eye teeth on a husky problem in addition, we can try something along the same line in subtraction:

3140

—2658

Again, we will repeat the problem at the top of each left hand page.

We begin attacking the problem (mathematicians always "attack" problems, as if they were fire-breathing dragons) by clearing the abacus.

The next step is to feed in the larger figure on top, which you can call the *minuend*, if you're technically minded. Since this is only a four-digit number, you should be able to do it in less than four seconds. Ready? Go! When you have finished, this is what your abacus should look like:

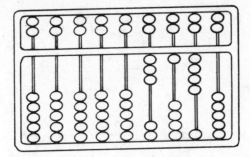

Now, we begin to feed *out* the smaller number, the *subtrahend*. We do this beginning with the 8 on the right and ending with the 2 on the left, taking each digit at a time. All the work should be done on the abacus; don't look at the page except to note the subtrahend digits.

Our first job, as we see, is to take 8 from 0. Here is a snag before we can even get started. How can you take eight from nothing?

It sounds hard, but there's really nothing to it. We think of the same "bright idea" we used in addition. Only here, since subtraction is the opposite of addition, we work it the opposite way. Taking one of the earth beads at the crossbar in column *two* out of the calculation by sending it to the bottom of the frame is exactly like subtracting 10 from column one. That is true because every earth bead in the second column is worth ten in the first, as we know. However, we don't want to subtract 10, we only want to subtract 8. So, with the same motion with which we remove the earth bead from column two, we bring two earth beads up to the crossbar in column one:

$$3140$$
$$-2658$$

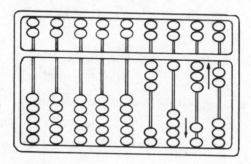

Our next job is to take 5 out of the second column. But this column now has only three beads at the crossbar. How do we subtract 5 from 3? By using the same "bright idea": we flip the one earth bead of column *three* down to the bottom of the frame to subtract 10 from column two, and at the same time bring one heaven bead in column two down to the crossbar to add 5:

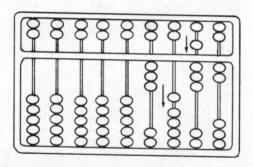

We now move on to the next digit of the subtrahend, which is a 6. Since this is the third digit from the right, it must be taken out of the third abacus column. But this column, as we see from the drawing just above, has no beads at the crossbar. To feed 6 out of this column, then, we subtract 10 and add 4. And we do this by sending one of the earth beads in column *four* down to the bottom of the frame while bringing four earth beads in column three up to the crossbar.

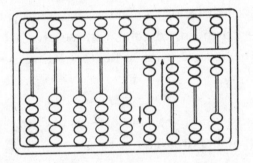

Our last job is to feed the 2 of the subtrahend out of column four of the abacus. Since there are exactly two beads left at the crossbar in this column, we simply sweep them down to the bottom of the frame.

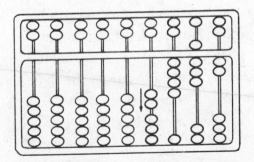

$$\begin{array}{r} 3140 \\ -2658 \\ \hline \end{array}$$

With all the digits of the subtrahend taken care of, we see that the answer is given in the picture just above. How do we translate that into figures? Well, the two earth beads in column one give us a 2, the one heaven and three earth beads in column two add up to 8, and the four earth beads in column three give a total of 4. Writing these digits from right to left, we see the answer is

482

which reads, since numbers are always read from left to right, four hundred eighty-two.

In the problem we just did, whenever we had to subtract a larger number from a smaller one in a column, we always had earth beads at the crossbar in the column just next door. Everything would be rosy if things always turned out that way. But they don't. Suppose, for example, you had a ten-dollar bill and you bought a candy bar for six cents. How much change could you expect?

This is plainly a problem in subtraction, because it means taking six cents away from the thousand pennies in a ten-dollar bill and seeing what we have left. In figures, the problem would be

$$\begin{array}{r} \$10.00 \\ -.06 \\ \hline \end{array}$$

We can see how it's done with pencil and paper. But how is it done on the abacus?

We begin this problem as we would begin any other in subtraction. After clearing the abacus, we feed in the minuend. This consists of three 0's and a 1. Since this 1 is the fourth digit from the right, we feed it into the abacus by bringing one earth bead in column four to the crossbar:

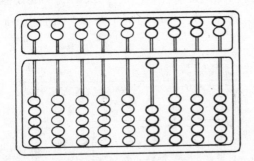

Notice that no beads have been set at the crossbar in the first, second, and third columns. This is only natural since the three 0's at the right of the minuend amount to nothing, but simply hold their positions open.

Now, we would like to subtract 6 from the first column. But there aren't any beads at the crossbar in this column, nor are there any at the crossbar in columns two or three. What to do? Before reading on to see how this problem

$10.00
−   .06

is solved, take a few minutes to think how you would do it, then read ahead to see if you're right.

Have you thought about it? Here's how it's done:

We know that the single earth bead in column four is worth ten in column three. So, let's make the exchange by bringing the bead in column four down to the bottom of the frame, at the same time pinching one heaven and five earth beads to the crossbar in the third column, like this:

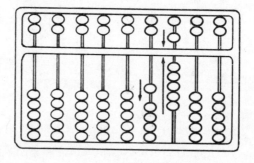

Now we're beginning to get some ammunition to attack our problem with. We can get some more by making a second exchange—by trading one earth bead of column three for one heaven and five earth beads in column two:

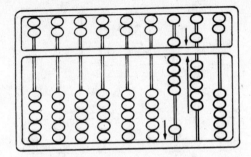

We now have beads in column two to work with, which means that we could solve the problem right now. But just to see what happens, let's go one step further and exchange one of the earth beads in the second column for one heaven and five earth beads in the first:

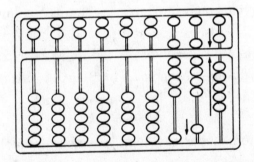

What we have in this last picture is exactly the same, believe it or not, as what we had in the abacus at the start of this problem. The difference is that now we can make the subtraction very easily by taking the 6 out of the first column. That we do by sweeping the heaven bead at the crossbar to the top of the frame and one earth bead to the bottom:

$10.00
−   .06

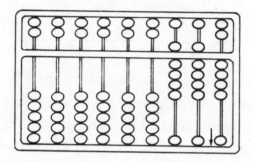

And we can read our answer. The first column, with four earth beads, is a 4; the second column, with one heaven and four earth beads, adds up to 9; the third column, with one heaven and four earth beads also adds up to 9. That gives us, in dollars and cents,

$9.94

If the storekeeper gave you less than nine dollars and ninety-four cents, rework it for him on the abacus.

We promised you that things would get rough in this chapter, and they will. We warmed up on addition, we got a bit warmer on subtraction, and now we are ready for multiplication, where things really get hot.

Let's try a fairly easy problem, just to keep from getting too badly scorched right away:

$$68$$
$$\times 7$$

It is good abacus practice, when you have numbers of two or more digits to multiply, to feed both these numbers into the abacus. In that way, you won't forget the original numbers you are multiplying. Furthermore, it is best to feed the two numbers into the columns at the left of the abacus. That will leave the columns at the right open for the calculation.

So, after clearing the abacus, we feed the 68 into the last two abacus columns on the left, and then, leaving the next column vacant, feed in the 7, like this:

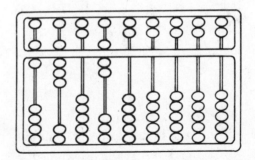

The purpose of having the vacant column is to keep the two numbers separated. If they were fed into the three last columns, it would be hard to tell whether we were multiplying 68 by 7 or 6 by 87. Looking at the drawing just above, we can see that 68 is the *multiplicand* —the name given to the number being multiplied—and that 7 is the *multiplier*.

In multiplication, as in anything else on the abacus, we work from right to left. Since the 8 appears on the right of the multiplicand, we multiply it by the 7 first. Now, as every good abacist should know, 7 times 8 are 56. Since our calculations are made in the columns to the right, we feed the 56 into the first two columns there:

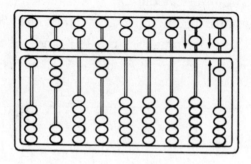

The 6 is put into the first column by pinching one earth bead and one heaven bead at the crossbar; the 5 is simply one heaven bead at the crossbar in the second column.

We now multiply the other number in the multiplicand, the 6, by the multiplier 7. Now, 7 times 6 are 42. Where do we feed these digits into the abacus? Since the 6 is the second digit from the right in the multiplicand, the 2 of the 42 goes into the second column from the right of the abacus. And, if the 2 goes into the second column, the 4 would naturally have to go into the third:

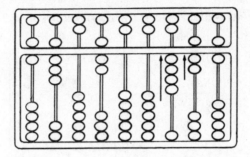

Since the 7 has now been used to multiply both digits in the multiplicand, the problem is finished. And we can read the answer from the first three columns. Column one has a heaven bead and an earth bead at the crossbar for a total of 6. Column two has a heaven bead and two earth beads for a total of 7. Finally, column three has four earth beads for a total of 4, and our answer is written as:

476

and is read as four hundred seventy-six.

Since that problem wasn't too much trouble, let's try something just a little harder. This time our multiplicand and multiplier will have two digits each:

75
×96

In this example, 75 is the multiplicand—since it is on top—while 96 is the multiplier. Actually, it doesn't matter which is which, for 75 × 96 is exactly the same as 96 × 75. Anyway, we feed both numbers into the left end columns of the abacus, with a vacant column be-tween them—after clearing it, of course:

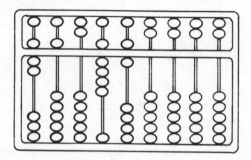

As we did before, we begin with the digits to the extreme right. Multiplying the 5 of the multiplicand by the 6 of the multiplier, we get 30. The 0 goes into the first column of the abacus. But, as we already know, the abacus is concerned with 0 only as an empty column, so the first column remains as is. However, a 3 should be fed in alongside it, so we lift three of the earth beads in column two to the crossbar:

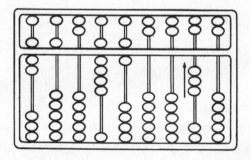

We now multiply the 7 of the multiplicand by the 6 of the multiplier. The result of that is 42. Since the 7 is the second digit from the right of the multiplicand, it is really 70 since 0 fills the position of the 5 "dropped"; therefore the 2 of the 42 must be fed into the second column of the abacus and the 4 alongside it into the third, leaving the first column open for the 0. The 2 is fed into the second column simply by dropping one of the heaven beads in this column to the crossbar, at the same time sending the three earth beads already at the crossbar down to the bottom of the frame. This amounts to adding 5 and subtracting 3, and that means adding 2 to the second column. As for feeding the 4 into column three, that is done very easily by raising four earth beads in this column to the crossbar.

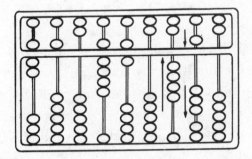

The 6 of the 96 multiplier has now done its part. Having multiplied both digits of the multiplicand, it is of no

further use. So we return its one heaven bead to the top
of the frame and its one earth bead to the bottom.

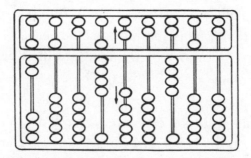

We do this for two reasons. In the first place, it leaves
the abacus looking less cluttered with beads, and that
always helps. In the second, it leaves more room for the
calculation going on in the right-hand end of the abacus.
For this particular example, we don't need the extra
room; but if you were to multiply numbers of more than
two digits, the extra calculating space would certainly
be useful.

Now, all that remains of our problem is to multiply
the 9 of the multiplier still left in the abacus by both
digits of the multiplicand. We first multiply 9 by the 5
to get 45. Since 9 is the second digit of 96 (counting from
the right, of course) the 5 of the 45 is fed into the second
column while the 4 goes into the third. To feed in the 5,
we could raise the five earth beads of column two to the
crossbar. But that would clutter the column with beads;
and it is a good idea, as we have already mentioned, to
keep from doing that. A neater way is to add 10 and sub-

tract 5—that is, to raise the one earth bead still at the bottom of the frame in column *three* to the crossbar, and send the heaven bead already at the crossbar in column two up to the top of the frame.

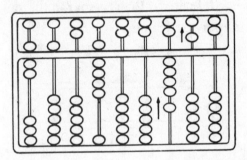

But we still have to feed a 4 into the third column. We do it by lowering one of the heaven beads in this column to the crossbar; at the same time, we slide one of the earth beads in this same third column to the bottom of the frame. In this way, we add 4 by adding 5 and subtracting 1.

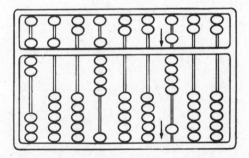

The final multiplication is made between the 9 of the multiplier and the 7 of the multiplicand. This gives us 63, the 3 to be fed into the third column while the 6 is fed into the fourth. To feed in the 3, we raise one earth bead in the *fourth* column to the crossbar. And, in the same motion, we return the one heaven bead at the crossbar in column three to the top of the frame, and send two of the earth beads at the crossbar in the same column to the bottom of the frame:

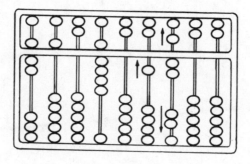

Finally, we feed the 6 of our 63 into column four by pinching one heaven and one earth bead together at the crossbar:

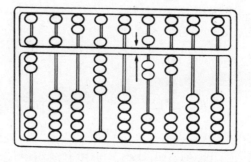

And now we can read off our answer. There are no beads at the crossbar in columns one and two, which means that the two digits at the right of our answer must be 00. The third column has two earth beads at the crossbar, so the third digit from the right in the answer must be a 2. The fourth column, with one heaven bead and two earth beads, has a total of 7. And that means that the fourth digit from the right in the answer must be that number. Putting this all together, we can see that the result of our multiplication is

7200

On a nine-column abacus, it's not easy to multiply figures of three or more digits each. There just isn't enough room for the calculation with the multiplier and multiplicand also taking up space. And that is where the larger abaci (some have as many as twenty-seven columns!) are useful. Still, there are times when even big figures can be handled by a small abacus.

For example, would you need an abacus of more than nine columns to multiply 2,000,000 by 74,000? Certainly not. Since all those zeros do not change the positions of the abacus beads on their rods, they can be kept in your memory while you multiply 74 by 2. You then take the six zeros in the first number and the three zeros in the second out of your memory, and tack them onto the result. Let's do this multiplication.

After the abacus is cleared, we feed 74 into the last two columns on the left, leave the next column blank, then feed the 2 into the fourth column from the end:

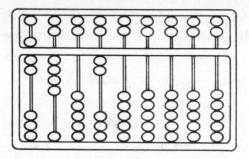

Ready? Here we go. We first multiply the 2 by the 4 of the multiplicand. That gives us 8. We feed this 8 into the first column with one heaven and three earth beads:

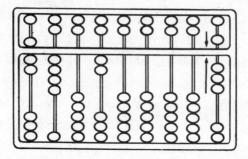

We now multiply the 2 by the 7 of the multiplicand to get 14. The 4 is fed into the second column and the 1 into the third:

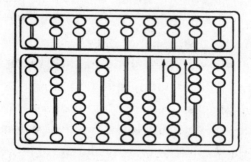

And, as we see from the final pattern of the beads, the result of the multiplication is 148. Now we tack on the nine zeros, and the answer is

$$148,000,000,000$$

which, of course, is read as one hundred forty-eight billion.

Division is tough, too. But that shouldn't frighten you. Its dragons are like those the Chinese use in their celebrations. They look fierce, make horrible noises, and spout steam from their nostrils; but when you take a good look at them, you find they are only flimsy paper draped over a few playful boys.

Actually, division is the opposite of multiplication. And we'll see that in the two examples worked out here.

For a start, suppose we try a fairly easy example:

$$12\overline{)144}$$

Like multiplication, division has its technical terms. The larger number being divided up (the 144 in our example) is called the *dividend*. The smaller number doing the dividing (the 12) is the *divisor*. A fancy word for the result is the *quotient*, but nobody is likely to get upset if you call it simply the "answer."

As we did with multiplication, we feed the divisor and dividend into the columns at the left-hand end of the abacus with an empty column between them. The divisor is fed into the last two columns, and the dividend into the next three—except for the empty column, that is:

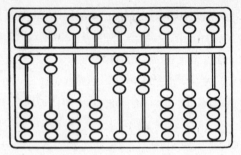

The columns on the right are used for the quotient.

For a moment, let's forget the 4 at the right of the dividend, and think how many times the 12 of the divisor could go into 14. Since 14 isn't much bigger than 12, the chances are that 12 can't go into 14 more than once; that is, there can't be more than one 12 in 14. The first digit from the left of our quotient, then, should be a 1. Where do we feed this 1?

We can find the answer to this question if we realize that we are really thinking of our dividend as 140, since 0 fills the position of the 4 "dropped," and are dividing it by 12. We can guess that the answer will be more than 10, so we feed the 1 into the *second* column of the abacus. We leave the first column open for the next digit of the quotient—when we find it, of course:

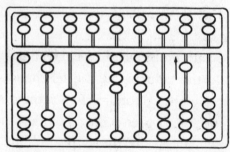

To see how much of the 14 we have left after its division by 12, we multiply both numbers of the divisor in turn by the 1 of our quotient, then subtract the result from the first two digits of the dividend. The first digit of the divisor is a 1; multiplying it by the 1 of the quotient gives a 1. We subtract this 1 from the first dividend digit by returning to the bottom of the frame the one earth bead to the right of the empty column between dividend and divisor:

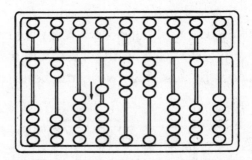

That takes care of the 1 in the divisor. Now we tackle the 2. We multiply it by the 1 of the quotient, which gives us a 2, and subtract this 2 from the next column of dividend beads. This column, as we can see from the drawing, contains four earth beads. To make our subtraction, we return two of these beads to the bottom of the frame. And what we now have is

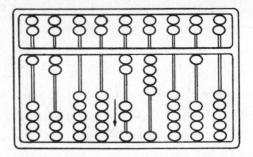

We still have our original divisor of 12 at the extreme left of the abacus. But all that remains of the dividend is 24—two earth beads in one column, and four earth beads in the next. How many times will our divisor of 12 go into this 24? We'll guess two; 24 seems big enough to accommodate two 12's, but doesn't seem big enough for three. In any case, we feed the 2 into the first column of the abacus by bringing two earth beads up to the crossbar:

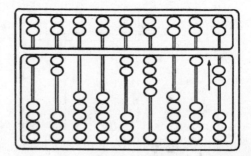

To see if our guess of 2 was correct, we multiply the two digits of the divisor by it and then subtract the result of that multiplication from what remains of the dividend. The first divisor digit is a 1. This, multiplied by the 2 of the quotient gives 2. We subtract this from the 2 of the dividend by sending the two earth beads at the crossbar in the fifth column down to the bottom of the frame:

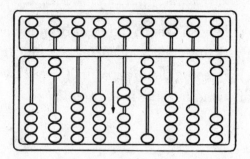

Now we take care of the 2 of the divisor. We multiply it by the 2 of the quotient, and that gives us 4. To subtract this 4 from the dividend, we remove the four earth beads left in the one remaining dividend column.

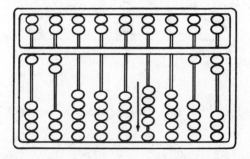

What has happened to our dividend? It has melted away, and that tells us that our problem is finished. What is the answer—or quotient, if you want to be technical? As you can see from the two columns at the right, with two earth beads in column one and one earth bead in column two, the quotient must be

12

If you want to try your hand at another division, take a look at this one:

$$23\,\overline{)713}$$

This problem isn't much different from the one before, and is solved the same way. After clearing the abacus, we begin by feeding the divisor into the last two columns and the dividend into the next three:

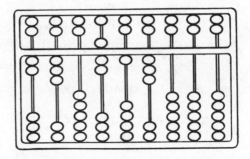

with a vacant column (no beads at the crossbar) between.

We now ask ourselves, how many times will the 2 of the divisor go into the 7 of the dividend? Quick as a flash we get the answer, 3 times. So we feed three earth beads into the second column, leaving the first open, as we did before (because *700* can be divided by *20* about 30 times, and therefore the three earth beads we feed in mean three tens and in the second column each earth bead represents ten units):

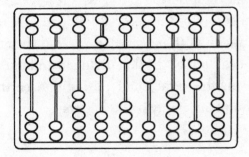

This three is part of the quotient. We multiply the first digit of the divisor by it to get 6. Subtracting this 6 from the first digit of the dividend, we feed one heaven and one earth bead out of the first dividend column by slipping the heaven bead to the top of the frame and the earth bead to the bottom:

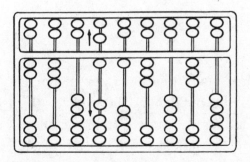

Now we multiply the second divisor digit, the 3, by our quotient digit of 3. That gives a result of 9, and this we must feed out of the second dividend column. We do it by subtracting 10 and adding 1. That means feeding one earth bead *out* of the first dividend column (the sixth column from the right) and raising one earth bead in the column at its right to the crossbar:

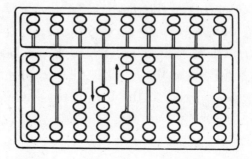

Now, how many times can our divisor go into what remains of the dividend? The digit to the left of the divisor is a 2, as we can see from the abacus, and so is the digit to the left of the dividend. The chances are, then, that the divisor goes into the dividend one time. So, we bring an earth bead in the first column up to the crossbar:

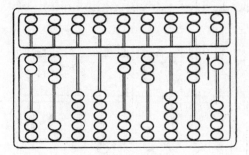

We multiply this 1 in our quotient by the 2 of the divisor. The result of this operation is, of course, a 2. We feed this out of the dividend (the fifth column from the right of the abacus) by flipping the two remaining beads at the crossbar down to the bottom of the frame:

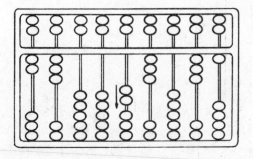

We now multiply the 1 of the quotient by the 3 of the divisor to get 3. That means sweeping the remaining three beads of the dividend down to the bottom of the frame:

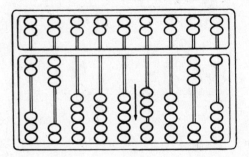

Our dividend thus disappears, and the problem is solved. What is the quotient? We can read it from the first two columns of the abacus:

31

. . . Seven looked up and said "That's right, Five! Always lay the blame on others!"

"You'd better not talk!" said Five. "I heard the Queen say only yesterday you deserved to be beheaded."

10. The private lives of numbers

The examples of the previous chapter, and, in fact, all problems in mathematics, are like stories. At the beginning, two or more numbers are introduced. They may be even-tempered and agreeable, like 2 and 4, or perhaps they're cross and pigheaded, like 7 and 13. But nice or nasty, they go about their business, have their adventures, fall in love or quarrel angrily. And just when their lives seem hopelessly botched, a little clearheaded thinking sends their troubles scattering, and the story ends happily: the answer comes out right.

Not always, of course. Sometimes the problem goes wrong in the middle; then the numbers have to be put through their paces again. And there are times, too, when the problem can't be solved at all. There just isn't any answer. Then the story doesn't have a happy ending. It's pretty sad, in fact—especially for the mathematician.

But no matter how the story ends, you have to keep track of the characters in it. Otherwise it becomes dull instead of being the adventure tale it should be. I'm sure you have read books where you found something like this:

"Hello, Tom," said Gerald when he met his friend Sanders at the football game, "did you tell Mary about the trouble Richard was having with Godfrey Pumper-, nickel?"

Well, after some hard thinking, you may remember who Richard and Mary are, and even that Tom's last name is Sanders. But you can't for the life of you recall anybody named Godfrey Pumpernickel. And that stops you cold. Right then, you throw aside the book and start wondering what else you can do with yourself. Now, that might be a mistake. Maybe, if you stuck with it and sorted all those characters out properly, that story would have been the best thing you ever read.

It's the same way with this book. If you find your head swimming with all the numbers floating around here, there, and yonder, give it a shake and start again. You'll find it worth while. Numbers are fascinating things, especially when you have something like the abacus with which to tame them. And mathematics is full of delightful surprises. It is not always sensible and clear; it has its dark, unsolved mysteries. There are problems which have been attacked again and again by the sharpest minds in history, but with no success. Perhaps you will be the hero who someday will find the answers to these riddles.